This book addresses key issues in social theory such as the basic structures of social life, the character of human activity, and the nature of individuality.

Drawing on the work of Wittgenstein, the author develops an account of social existence that argues that social practices are the fundamental phenomenon in social life. This approach, while respecting the multiplicity and heterogeneity emphasized in recent social thought, offers new insight into the social constitution of individuals, surpassing and critiquing the existing practice theories of Bourdieu, Giddens, and Lyotard.

The author thereby shows the relevance of Wittgenstein's work to a range of social theoretic issues to which it hitherto has not been applied.

The book will be of particular interest to social and continental philosophers, philosophers of the social sciences, a wide range of social theorists in sociology and political science, as well as some literary theorists.

SOCIAL PRACTICES

SOCIAL PRACTICES

A Wittgensteinian Approach to
Human Activity and the Social

THEODORE R. SCHATZKI

University of Kentucky

CAMBRIDGE
UNIVERSITY PRESS

Published by the Press Syndicate of the University of Cambridge
The Pitt Building, Trumpington Street, Cambridge CB2 1RP
40 West 20th Street, New York, NY 10011-4211, USA
10 Stamford Road, Oakleigh, Melbourne 3166, Australia

© Cambridge University Press 1996

First Published 1996

Printed in the United States of America

Library of Congress Cataloging-in-Publication Data
Schatzki, Theodore R.
Social practices : a Wittgensteinian approach to human activity
and the social / Theodore R. Schatzki.
p. cm.
Includes bibliographical references.
ISBN 0-521-56022-5 (hardcover)
1. Sociology – Philosophy. 2. Social sciences – Philosophy.
3. Wittgenstein, Ludwig, 1889–1951. I. Title.
HM51.S346 1996
301'.01 – dc20   95-42276
                                                       CIP

A catalog record for this book is available from the British Library

ISBN 0-521-56022-5 hardback

For my loving parents, Ginny and Stefan

Contents

Abbreviations

Wittgenstein's Texts

Unless otherwise noted, all references to Wittgenstein's texts employing the following abbreviations are to section (paragraph) number.

LW I   *Last Writings on the Philosophy of Psychology, Volume I: Preliminary Studies for Part II of Philosophical Investigations,* ed. G. H. von Wright and Heikki Nyman, tr. C. J. Luckhardt and M. A. E. Aue. Chicago, University of Chicago Press, 1990.

LW II   *Last Writings on the Philosophy of Psychology, Volume II: The Inner and the Outer,* ed. G. H. von Wright and Heikki Nyman, tr. C. G. Luckhardt and Maximilian A. E. Aue. Oxford, Blackwell, 1992.

PI   *Philosophical Investigations,* tr. G. E. M. Anscombe. New York, Macmillan, 1958.

OC   *On Certainty,* tr. Denis Paul and G. E. M. Anscombe. Oxford, Blackwell, 1977.

RFM   *Remarks on the Foundations of Mathematics,* rev. ed., ed. G. H. von Wright, R. Rhees, and G. E. M. Anscombe, tr. G.E.M. Anscombe. Cambridge, Mass., MIT Press, 1978.

RPP I   *Remarks on the Philosophy of Psychology,* vol. 1, ed. G. E. M. Anscombe and G. H. von Wright, tr. G. E. M. Anscombe. Oxford, Blackwell, 1980.

RPP II   *Remarks on the Philosophy of Psychology,* vol. 2, ed. G. H. von Wright and Heikki Nyman, tr. C. G. Luckhardt and M. A. E. Aue. Oxford, Blackwell, 1980.

Z   *Zettel,* ed. G. E. M. Anscombe and G. H. von Wright, tr. G. E. M. Anscombe. Berkeley, University of California Press, 1967.

Works of Giddens and Bourdieu

CP   Anthony Giddens, *Central Problems in Social Theory.* Berkeley, University of California Press 1979.

CS   Anthony Giddens, *The Constitution of Society.* Berkeley, University of California Press, 1984.

LP   Pierre Bourdieu, *The Logic of Practice,* tr. Richard Nice. Stanford, Calif., Stanford University Press, 1990.

OT   Pierre Bourdieu, *Outline of a Theory of Practice,* tr. Richard Nice. Cambridge, Cambridge University Press, 1976.

Preface and Acknowledgments

This book was a long time coming. A necessary detour in a previous manuscript through an individualizing appropriation of Heidegger's conception of *Dasein* and world delayed return to the encompassing context of practices announced in Wittgenstein. The current work attempts to retain insights of the detour within a practice framework.

The book is organized as follows. Chapter 1 describes my aim of developing a conception of practice that positions social practices as the central phenomenon in social life. It also locates this project in the swirling currents of contemporary social thought. The venture then commences in Chapter 2 with an account of mind/action developed as a systematizing interpretation of Wittgenstein's late remarks on mentality. Following this, in Chapter 3, an analysis of the social constitution of mind/action concludes by insinuating practices as the social context of this "process." In Chapter 4, I present my account of social practices, focusing on the configuration of activity and articulation of intelligibility that occur within them. Chapter 5 argues for the superiority of this account over those of Bourdieu, Giddens, and Lyotard. The book then concludes in Chapter 6, where social practices are revealed to be the central context of human coexistence and initial steps are taken toward conceiving the social field as a nexus of practices.

The manuscript was written while on fellowship in Norway and Germany. I would like to thank both the Council for International Exchange of Scholars and the United States–Norway Fulbright Foundation for Educational Exchange for the award of a Fulbright Research Award spent at the University of Bergen during the summer–fall of 1993; and the Alexander von Humboldt-Stiftung for award of a Humboldt research fellowship spent at the Universität Bielefeld from the late fall of 1993 to the fall of 1994. Like those preceding me, I can only extoll the support, generosity, and high-mindedness of these agencies.

In Bergen my hosts were chiefly Kjell Johannessen of the philosophy department and Claus Huitfeldt, director of the Wittgenstein Archives. I would like to thank both, especially Johannessen, for their acumen, conversations, and kindness. A much appreciated office was also provided at the archives, a splendid intellectual and scholarly setting in which to ponder Wittgenstein's texts. My host in Bielefeld was Eike von Savigny. Professor von Savigny provided exceptional intellectual and social support throughout the year in Bielefeld. I would further like to thank the excellent philosophy department there for its hospitality as well as for the use of one of its offices.

Above all, thanks go to Professors Johannessen and von Savigny for their penetrating and insightful comments on the manuscript. Their input has certainly improved the text. In Bergen I benefited, in addition, from the sagacity of Tore Nordenstam. In Germany, parts of the manuscript were presented in various seminars, and I would like to thank participants in those settings for their comments and conversations, in particular Rudiger Bittner, Karin Knorr-Cetina, Stefan Hirschauer, and Albrecht Wellmer. An older and deeper intellectual dept is due Hubert L. Dreyfus and Charles Taylor, who in their teaching and writing ingeniously insist and earlier focused my attention upon the significance of practices.

Finally, I would like to thank Nora Moosnick for her love and that trip to St. Anton; Aashild Grana for her apartment in Bergen, where the most fateful decisions were wrestled out; and Dean Richard C. Edwards of the College of Arts and Sciences at the University of Kentucky for an extended leave.

1

The Emergence of Practice

When considering the nature of social life, social theory has always availed itself of two master concepts, those of totality (whole) and the individual. As often noted, this penchant debuted in the *Republic*, where Plato tied his discussion of justice, education, psychology, art, and social existence generally to an analogy between the individual and society as a whole. Although this analogy remained a powerful metaphor and organizing principle in subsequent thought about social life, its two terms disengaged in the modern era and have often been developed independently. Most modern social theorists have promoted either the individual or the social whole as the fundamental ontological phenomenon, and sought to analyze the other once equally insistent concept on its terms. This bifurcation solidified in the late nineteenth and twentieth centuries with the association of the divergent political ideals and agendas of liberalism and socialism with the two paths of social ontological conceptualization.[1] The opposition between these two ontological-political schools of thought has waned today. But the division in ontological proclivities still deeply rends social thought.

In the meantime, challenges have arisen to the integrity of the concepts defining the ontological front. Various currents in contemporary thought join in disparaging both the supposition of wholeness in social life and its construction in social thought. Many of these currents also undermine the unity and integrity of the individual subject and thereby problematize any construal of social existence as simply interrelations among individuals. Promulgating the charge on the two fronts simultaneously are the divers schools of thought whose convergence is emphatically but problematically marked by the term "postmodern," along with a collection of thinkers I will call "practice theorists." In stripping social thought of the two axes around which it has traditionally structured its ideas and in part prosecuted its large-scale internecine struggles, these movements have led some theorists to focus exclusively on limited interventions in selected theoretical and practical contexts. Others, unwilling to abandon general or global theorizing, have sought new ontological organizing concepts and points of departure for the theorization of social life. These new starts promise not only fresh conceptualizations of the nature of sociality, but also deeper understandings of individuals and of whatever wholes can be discerned in social life. In this opening chapter, I situate my project within the landscape carved by the traditions and doctrines of individualism, "wholism," postmodernism, and practice theory. I begin by reviewing and illustrating recent challenges to the two traditional master concepts.

The concept of a social totality is the concept of a social whole that is something more than its parts. This means, first, that a totality has an existence beyond that of its parts and a nature that transcends the properties of the amalgamation of these parts; and, second, that its existence and nature specify properties of and/or meanings for its parts. These properties and meanings usually pertain to the "place" the parts occupy in the whole. The existence, persistence, and development of the whole are also typically thought to be governed by principles that apply to the whole *qua* whole, thereby derivatively specifying the operations of the parts. Although almost any social institution or formation can be treated as a totality, perhaps the historically most important type of whole has been society. Beginning with Plato, and enjoying a renaissance in the nineteenth and twentieth centuries, the notion of society as a bounded and unified totality has guided theorists as diverse as G. W. F. Hegel, Karl Marx, Emile Durkheim, and, more recently, Bronislaw Malinowski, Louis Althusser, Talcott Parsons, and Niklas Luhmann. On their theories, social phenomena of lesser scope such as rituals, political institutions, families, and ideologies are assigned places as parts of society by reference to achievements necessary for the equilibrium or functioning of society; contributions to society's survival, fitness, persistence, or slow, directional evolution; or a series of overarching principles of freedom, reason, or value.

Contemporary opponents of the concept of social totality insist on the preeminence of the particular, local, and transitory in social existence. In their eyes, treating the intricate and complex tangle of phenomena that constitutes social life as neatly tied up in a system and governed by systemic principles neglects the contingent, shifting, and fragile relations among social phenomena that weave them into everchanging constellations. The point is not, at least usually, that these phenomena are autonomous and isolated, but instead that they constitute complex nexuses that do not add up to something beyond themselves. As Derek Gregory writes,

These writers do not, of course, deny the importance of the interdependencies which have become such a commonplace in the late twentieth-century world, and neither do they minimise the routine character of social reproduction nor the various powers which enclose our day-to-day routines. . . .

But they do object to the concept of totality . . . because it tacitly assumes that social life somehow adds up to (or "makes sense in terms of") a coherent system with its own superordinate logic.[2]

This emphasis in turn underpins the conviction that social theory is adequate to reality only if it registers in its concepts the tangled and unevenly propagating mosaic that is social existence. As a result, a bevy of theorists, from those exploring specific interconnections among different sectors of the social world to those tracing the fragmentation of images and ways of being in late twentieth-century Western life, has sought to spin out a conceptual grid equal to the newly acknowledged complexity and dispersion.

Zygmunt Bauman well captures the antitotality mood in a statement that, because of its clarity and trenchancy, deserves to be quoted at length and without comment:

What the theory of postmodernity must discard in the first place is the assumption of an "*organismic,*" equilibrated social totality it purports to model in Parsons-like style. . . . The sought theory must assume instead that the social condition it intends to model is essentially and perpetually *unequilibrated:* composed of elements with a degree of autonomy large enough to justify the view of totality as a kaleidoscopic – momentary and contingent – outcome of interaction. The orderly, structured nature of totality cannot be taken for granted; nor can its pseudo-representational construction be seen as the purpose of theoretical activity. The randomness of the global outcome of uncoordinated activities cannot be treated as a departure from the pattern which the totality strives to maintain; any pattern that may temporarily emerge out of the random movements of autonomous agents is as haphazard and unmotivated as the one that could emerge in its place or the one bound to replace it, if also for a time only. All order that can be found is a local, emergent, and transitory phenomenon; its nature can be best grasped by a metaphor of a whirlpool appearing in the flow of a river, retaining its shape only for a relatively brief period and only at the expense of incessant metabolism and constant renewal of content. . . . With the totality dissipated into a series of randomly emerging, shifting and evanescent islands of order. . . .[3]

I will return later in this chapter to the problematic nature of this statement.

Other theorists have developed social ontologies that portray social institutions, formations, and structures as arising out of or constituted by local and small-scale, but less random and autonomous, social entities. In his famous and controversial analysis of power, for instance, Michel Foucault depicts power as a web of "relations of force" among individuals. (Relations of force include those of control, domination, determination, formation, empowerment, and constraint.) These relations are local and shifting links that join particular people; and social formations rest upon while also being characterized by unstable and evolving chains and coagulations of such relations. In analyzing power vis-à-vis these larger entities, consequently,

the important thing is not to attempt some kind of deduction of power starting from its centre and aimed at the discovery of the extent to which it permeates into the base. . . . One must rather conduct an *ascending* analysis of power, starting, that is, from its infinitesimal mechanisms, which each have their own history, their own trajectory, their own techniques and tactics.[4]

Elsewhere Foucault describes history as a maze of events formed by myriad forces, a thicket of historical phenomena having numerous beginnings and complex descents.[5] For him, social life is a multi-stranded spectacle determined by a maze of heterogeneous and local forces, whose movement is ceaseless and future undetermined.

Anthony Giddens, meanwhile, has proclaimed the outdatedness of the concept of society as a bounded and unified whole. The recent renaissance of this grandest concept of totality in the arsenal of social thought was initially nourished in the middle of the previous century by the Hegelian (and subsequently Marxist) image of total social units that inaugurate and take the lead in successive historical epochs. Later in that century, the concept was sustained through an analogy with biological organisms. This analogy, in turn, underpinned late nineteenth- and twentieth-century functionalist visions of societies as wholes composed of interlocking and functionally defined parts, a vision that was absorbed into post–World War II theories of societies as systems. Giddens has developed a theory of the social world that builds up institutions and structures out of practices that are not defined in relation to any wholes which they might help form, but instead through the interlocking matrices of rules (and resources) governing them. He portrays social reality, correspondingly, as a mosaic of interpenetrating, interdependent, and shifting practices. Concepts such as society, if taken in their traditional sense and not reinterpreted, become obsolete as a result for the purpose of understanding the nature of social life. Giddens, accordingly, redescribes societies as intersections of multiple sets of recurring practices "which 'stand out' in bas-relief" from the total network of interlocking practices and are rarely cleanly demarcated in space and time.[6] Another example of reinterpretation, much in the spirit of both Foucault and Giddens, is found at the beginning of Michael Mann's *The Sources of Social Power*. Introducing the conceptual framework that organizes his study, the author writes: "Societies are not unitary. They are not social systems. They are not totalities. We can never find a single, bounded society in geographical or social space. . . . Societies are constituted of multiple overlapping and intersecting sociospatial networks of power."[7]

Of course, this *Stimmung* of fragmentation in no way reigns in social ontological thought alone. Common to "postmodern" developments in a wide range of disciplines and social realms is the replacement of totalities with constellations of particulars. As an example, let us recall the ascension in contemporary life of what has been called "local politics." This example is of particular interest in the present context because of the ontological presuppositions that animate this form of political activity.

The modifier "local" in local politics is intended in part to contrast this politics with that directed toward the dismantling of large-scale structures and systems, such as the state and the capitalist economy. The latter politics earns the appellation "totalizing" for two reasons. First, it seeks the "revolutionary" overthrow of some "total" economic and political structure, usually hoping to replace it with an alternative whole economy or government. Second, guided by a theory of society that attributes dominant control to the targeted formation over most, if not all, other social sectors, it gathers together under the aegis of the one struggle against the governing instance all social oppositions and movements that aim to alter particular aspects of the social fabric. Local politics breaks with either and usually both postulates.

In the place of one commanding formation, local politics recognizes a variety of shifting sources and structures of oppression, misery, and discontent and aims to alleviate specific sufferings through opposition to the particular formations responsible for them. An example is the struggle to create greater opportunities for women to engage in and advance in the workplace. Another is the provision of greater dignity for the mentally ill caught in the clutch of psychiatry and asylums. Coordinate with the recognition of diverse sources of oppression, such politics acknowledges that different groups suffer these different oppressions. "New social movements" in Western societies such as the women's movement, the gay and lesbian movements, and the Afro-centrist movement are all local struggles that embody these two recognitions.

No longer opposing a single and total economic or state formation, such social movements also refuse to be collected under the umbrella of one master conflict. For many decades, the battle against capitalism provided for some an overarching framework into which struggles against any particular social institution, for any particular group, and in any country were to be integrated. Even before the demise of the cold war opposition of the capitalist and communist blocks, the new social movements, together with a plethora of third-world liberation movements, had thrown off this yoke and achieved varying degrees of autonomy as oppositional fronts. This fragmentation of course begets new questions about the coexistence, compatibility, and possible relations among particularistic movements, with their diverse, sometimes mutually incompatible agendas. These questions have of late been made more urgent by the reassertion of long-suppressed ethnic groups whose pursuit of nationalistic aspirations fuels a "tribal politics" that only aggravates fragmentation. Negotiating the shifting cacophony of local and particular demands is today one of the central challenges of national and international politics.

Totality, in short, has become problematic today. As indicated, however, this is equally true of the other center to which social theorists have adverted in the modern age when conceptualizing the structure of social existence. The tradition of individualism – or, as Parsons dubbed it, "utilitarianism" – has sought to comprehend the social world as consisting somehow in interrelations among individuals alone. In this century, such schools of thought as game theory, neoclassical economics, methodological individualism (e.g., Karl Popper), symbolic interactionism, and most versions of ethnomethodology have taken this path. They all give theoretical pride of place to the actions, strategies, mental states, and rationality of individuals, the cooperation, negotiations, and agreements reached among individuals, the rules, norms, and threats governing people's behavior, and the unintentional consequences of behavior that often extend beyond actors' purview.

In political philosophy, the theories most closely associated with individualist ontology, the liberal and natural law traditions, have of late been fiercely criticized for supposing that individual subjects are psychologically integral independently of their participation in social institutions and practices. The "communitarian" onslaught against this supposition maintains that human beings possess identities, needs, primary desires, and conceptions of the good only within the context of such social phenomena. This assault challenges the ontological primacy of the individual, for if the constitution of psyches and hence individuals systematically presupposes social institutions and structures, it is not obvious that these social phenomena can be construed simply as relations among individuals.[8] Of course, it is questionable whether the great representatives of these political traditions, for example, Hobbes and Locke, made the targeted supposition. The communitarian critique nonetheless challenges any contemporary construal of society as *nothing* but individuals and their relations.

Poststructuralists have mounted challenges to the ontological primacy of the individual similar to those advanced by communitarianism. An example is the Lacanian thesis, further developed by Julia Kristeva, that a human being becomes a self-conscious and intentional subject only by "entering" into signification – that is, by mastering the signifier/signified structure of language, whereby it becomes capable of reflexive predication (intentional mental and behavioral activity).[9] The signifier/signified structure of language (*la langue*) consists of two contingently shifting realms of modulating and unstable "differences," one defining individual signifiers (words, symbols), the other individual signifieds (meanings, concepts). Acts of speaking and writing (*la parole*), in combining signifiers and signifieds into signs with which something is said about something, relate these two nondiscrete realms in contingent and shifting ways. Like the social practices and institutions of the communitarians, the realms of differences that on this theory constitute individuals do not consist simply in relations among individuals. Of course, it is unclear exactly what is "social" about these realms, as opposed, say, to the speech acts of individuals.

Poststructuralist theories have also raised criticisms of the integrity of the individual that are much more biting than the thesis of social constitution. These criticisms largely abstract from the question of social constitution and contend that subjects, regardless of how they come to be, are internally fragmented and lack the unity traditionally associated with the notion of an individual. This line of thinking was inaugurated by Nietzsche's suggestion that the mind is a community of agencies, along with Freud's parallel diremption of the mind into parts, which the conscious ego is in principle unable either to control or fully to comprehend. Chantal Mouffe's work presents an especially clear and useful (for my project) contemporary example.

Like other writers, she denies that (personal) identity is best considered an inherent property of a subject that is always already present whenever there is experience or action. A paradigm example of the rejected conception is Descartes's assumption, made explicit by Kant, that thoughts and experiences are had and linked by a subject that is *eo ipso* an I. Identity in the sense of I-ness is not an inherent property of a thing or substance called the subject. It is instead a social construction, an achievement realized only through the incorporation of human beings into the institutions and structures of social life. This holds true, however, not only of this now traditional modern notion of identity, being an I, but also of the more recent and more contentfull notion of being a particular someone. A person's identity in this sense is who she is. Mouffe pushes the thesis of social constitution in the direction of subject fragmentation by conceiving the acquisition of identity in the sense of who one is as the assumption of "subject positions." Subject positions are determinations, or identifications, that enter into and help compose who people are. Examples are father, Jew, acrobat, type A personality, and overwrought. Such positions are made available to people by the practices in which they participate. As this short list suggests, moreover, the set of subject positions overlaps with but differs from that of roles. While some subject positions (e.g., mother, baker, professor) are also roles, others are not (e.g., Catholic, Norwegian, and African American); and not all the roles a person occupies are components of her identity. Who a person is consists in the particular ensemble of subject positions she assumes in participating in various social arenas. This ensemble is woven from the possible positions offered to her by practices in these arenas. And it is woven around certain determinations called "nodal points" that form the core of who she is at a given moment. This mélange is unstable not only because the nodal points and constitutive mix can and do evolve, but also because there can be no presumption that a given identity amalgam is coherent. The identity of the socially constituted subject is thus precarious and unstable.[10] There is, furthermore, nothing more to identity to anchor or otherwise steady it.

An important consequence of this conception of identity is that many allegedly shared general identities are illusory. Subject positions are constituted within particular practices. This means that they usually have a specificity germane to the practices involved. For instance, there may be no general identity "woman" shared by all individuals that qualify as women. For there may be no universal practices in which all such individuals participate (or features common to practices in which each participates), through which they all could assume the same undifferentiated determination. Such individuals are of course women, but this biological fact guarantees neither that "woman" is a subject position nor that being a woman is a significant component of their identities. Each woman assumes an amalgam of nodal points and subject positions defining who she is (as a woman), the particular ensemble depending on circumstances and on the particular social conditions and practices that encompass her. This means that nationality, ethnicity, religion, class, and the like almost always more strongly compose her identity that does the mere fact that she, like others, is a woman.[11] So the identities of women are manifold and disparate, and there is no common position "woman" shared among them. This implies, in turn, that there is no single, unified "woman's" movement relevant to all women, but a plethora of interrelated women's movements, each taking up the cause of the collection of women who share a particular subordinated subject position (such as black woman, working woman, Catholic woman, lesbian, and so on).

The fragmentation and destabilization of the subject in contemporary poststructuralist theory dovetails with the idea propagated in many prominent twentieth-century schools of thought, that the individual human subject is neither given, foundational, nor in charge of human action and the processes of meaning and significance. These schools range from structuralism, which disengaged signification from the operations of the subject and treated the subject as derivative from abstract structures of difference, to behaviorism, which did away entirely with the subject, which it denounced as a Cartesian "ghost in the machine," to cognitive psychology, some versions of which treat subjects and minds as nothing more than, or at best epiphenomena of, the brain. Of course, these and other movements conceive of the subject, mind, and/or individual differently and not always inimically to individualist social ontology. Enough, however, do challenge this ontology as to render problematic any attempt to conceive social life as fundamentally nothing but interrelations among acting and experiencing subjects. This conclusion does not deny that some social phenomena, in particular certain economic and political ones, can be reasonably modeled as nexuses of individual agents (e.g., in game theory). Modelers of specific states of affairs need not trouble themselves about the origin and constitution of the individuals they depict. A social ontology that seeks to grasp the fundamentals of social reality, however, must seek an alternative starting point.

Thus, waves of contemporary theory caution against theorizing the structure of social life beginning from the concepts of either totality or the individual. Of course, not all social theorists accede to current attempts to fragment social totalities or to destabilize the subject-individual and treat it as a derivative phenomenon. Such totality-wielding theories as Marxism, systems theory, and functionalism continue to enjoy advocates today, as do especially such individualisms as neoclassical economics, methodological individualism, symbolic interactionism, and game theory (though proponents of individualisms usually do not promote them as general social ontologies). There is, nonetheless, tremendous power in the accumulated force of these criticisms. This makes it intellectually incumbent upon proponents of the two master concepts to answer and attempt to neutralize these assaults, and requires partisans of these critiques to articulate them further and embark on alternative ontological perspectives.

My effort in this book to develop an alternative starting point only in part reflects the merits of recent disparagements of totality and individuality. It is more crucially rooted in the fact that the writings of the philosophers with whom I am most in sympathy clash with the foundational utility of these two concepts, while also motivating an alternate course. This means that I do not necessarily advocate the foregoing antitotality and antiindividuality theses. I sympathize with their general thrust in assailing the two master concepts, and will appropriate several in seeking to deepen new ways of thinking. But not all these ideas are compatible with the position to be defended and the arguments for it, let alone with one another. The main reasons for describing these prominent features of the current social theoretical conjuncture were, first, to sketch a part of the theoretical context out of which my own project arises and, second, to guide the reader in locating the ideas developed in subsequent chapters.

Some thinkers impressed with arguments against totalities have come to oppose all global or general accounts of social life, even those of particular sectors thereof. I do not dispute that the dissipation of totality counsels the propriety of diminishing expectations for the defendable fruits of general theorizing. It is illusory to seek theories that tie together all aspects and sectors of social life into a coherent, systematically interrelated package. But the deflation of this fancy does not entail the advisability of abandoning the search for general and fundamental features of social life (or of large-scale sectors thereof). For there is no reason to think that social life can exhibit such features only if it is a totality. Nexuses of local phenomena interrelated in diverse shifting and contingent ways can also reveal such characteristics to the theoretical gaze. Of course, general theory, as its opponents intone, must vigilantly guard against trivialities dressed up as general/fundamental truths. But there is no a priori reason why general and fundamental features of complex nexuses must be trifling and plain. In this regard, each account must be judged on its own merits.

Generally described, consequently, my project aims to develop key elements of a general conception of social life that is equal to the interwoven complexity and lack of totality emphasized by recent writers and underwrites better understanding of the social constitution of the individual. This project thus has the dual aims of illuminating the character and import of the social and of casting further doubt upon the foundational utility of the traditional ontological choices.

It is important to stress that the laudatory stance adopted in previous paragraphs toward particularistic ontologies that acknowledge complexity, contingency, and dispersion is not matched by advocacy of the relativistic epistemologies often (but often wrongly) associated with the theorists promulgating these ontologies. Such theorists usually contend that ideas and theories arise out of, and are beholden to, cultural, sociohistorical, disciplinary, gender, and biographical contexts. This characterization of thought and cognition is surely correct (although these theorists are hardly the first to espy this contextualization). And, as some writers have suggested, this situation requires a theoretician to examine where he or she is coming from and to present (as I have been doing) the formative contexts of his or her ideas. But it in no way licenses the relativistic dismissal of cognition, the claim that the dependency, conditionality, yes relativity of thought to context implies that we can not correctly grasp how things are, or at least grasp them better than before. It simply informs us of the inadequacy of previous self-conceptions of the nature of cognition and its achievements. An example is the thesis that the apparatus of cognition functions correctly when it acts like a mirror of the world, introducing nothing of itself into its products.[12] Acknowledging, therefore, the contextuality of thought should not hamper the attempt to theorize universal (though not necessary) features of social life. If this book's attempt to do so is inadequate, parochial, or skewed, then, I believe, this is so less because of sociohistorical embeddedness than because of unsuccessful theorization, abstraction, and conceptualization. As Jürgen Habermas has urged, the embeddedness of thought counsels not that we abandon the search for generality, but that we be conscious of the fallibility of our attempts.

Disengaging nominalist ontologies from relativistic epistemologies can also be understood, but only at the present theoretical moment, as the establishment of "postmodern" theories of social life as competitors to wholistic and individualistic theories of it. Commentators who both emphasize the relativistic conceptions of cognition and knowledge associated with the thinkers most often labeled "postmodern" and neglect these thinkers' accounts of the nature of social life sometimes wrongly aver that postmodernism offers merely critical, deflationary, and "deconstructive" analyses and is unable to construct illuminating positive counterpoints.[13] I would argue that fecund strands of "postmodern" ontological theory have been gaining momentum during the past two or three decades. And by this now affirmative use of the term postmodern, or better "post-nineteenth-century-modern," to embrace a range of thinkers wider than that usually associated with it, I mean simply theory that has left behind the either/or of totality or individuality.

One of the most promising impulses beyond this either/or is practice theory. By this I mean a collection of accounts that promote practices as the fundamental social phenomenon. Such theorists as Pierre Bourdieu, Anthony Giddens, Jean-François Lyotard, Charles Taylor, and, to some extent, Ernesto Laclau and Chantal Mouffe agree that practices are not only pivotal objects of analysis in an account of contemporary Western society, but also the central social phenomenon by reference to which other social entities such as actions, institutions, and structures are to be understood.[14] That they specify a particular type of entity, namely, practices, as the principal constitutive element in social life, does not mean that they advance "totalizing" accounts. Nowhere do large-scale, unified totalities appear in their theories. Although differing greatly among themselves, their accounts present pluralistic and flexible pictures of the constitution of social life that generally oppose hypostatized unities, root order in local contexts, and/or successfully accommodate complexities, differences, and particularities.

Given the considerable differences among their analyses, the term "practice theory" can only designate the family of conceptions of practice they develop. These conceptions do, however, in my view share at least one important trait: the idea that practices are the site where understanding is structured and intelligibility (*Verständlichkeit* and *Bedeuten*) articulated (*gegliedert*).[15] These theorists disagree, of course, about the nature of understanding and intelligibility as well as the character of their structure and articulation. Moreover, most of them do not employ the terms "intelligibility" or "articulation." I believe, however, that we gain deeper insight into what is transpiring in their accounts by attributing this idea to them in common. It is further helpful to see them as concurring that what justifies treating practices as the fundamental component of social life is the fact that understanding/intelligibility is the basic ordering medium in social existence.

By designating practices the fundamental social phenomenon, these thinkers have placed it at the center of social theory. Missing from this roll call, however, is the thinker who perhaps focused more intensely than anyone else on how practices carry understanding and intelligibility, Ludwig Wittgenstein. It would be odd, of course, to add Wittgenstein to the list, since he did not construct theories about anything, let alone social ontological matters. Nonetheless, I believe that it is possible to develop an account of practices (toward which he gestures with the term "language-game") on the basis and background of his painstaking observations and descriptions. This account not only, in my opinion, offers an incisive analysis of practice, understanding, and intelligibility, but can be employed to further the attempts of the foregoing thinkers to position practices as the central phenomenon in the tangle that is human sociality.

What's more, this account offers a captivating analysis of the social nature of individuals. In Wittgenstein's hands, understanding and intelligibility structure not only the social realm, but also the domain of individual mind and action. Practices, in addition to being the elements and circuits forming the "flexible networks"[16] in which the social field consists, also (1) help institute which mental states and actions humans are and can be in and (2) are the contexts in which humans acquire the wherewithal to be in these states and to perform the actions that compose practices. By virtue of the understandings and intelligibilities they carry, practices are where the realms of sociality and individual mentality/activity are at once organized and linked. Both social order and individuality, in other words, result from practices.

Among practice theorists, such a position is anticipated by Laclau and Mouffe and especially by Bourdieu and Giddens. The latter two theorists have seen fully that the realms of mentality and sociality are jointly and coordinately constituted by one and the same sort of phenomenon. In Giddens's rendering, "social practices, biting into space and time, are . . . at the root of the constitution of both subject and social object."[17] I believe, however, that the Wittgensteinian version of this thesis is more comprehensive than theirs and that it identifies a broader basis for individuality and sociality. Wittgenstein's work thus harbors a seminal analysis of practices with significant implications for the constitution of the individual and the nature of sociality. Unfortunately, apart from Lyotard, David Bloor, and to a lesser extent Giddens and Laclau and Mouffe, social theorists have generally ignored his work.[18]

The aim of my project can thus be reformulated as developing an account of practice, based on and inspired by Wittgenstein's work, that forms the centerpiece of a conception of the social that is adequate to the complexity and lack of totality emphasized by recent writers and undergirds better understanding of the social constitution of the individual.

I add parenthetically that in speaking of the articulation of intelligibility, I do not imply any particular significance for language therein. As we shall see, language alone does not articulate intelligibility – bodily behavior and reactions also play an omnipresent and foundational role. Language is also unable to articulate fully the understandings and intelligibilities that permeate human life. Thus in saying that practices are both the most basic social phenomenon and the site where intelligibility is articulated, I do not suggest that social existence reduces to language in some fundamental way or that everything about practices can be expressed in language. I mention these matters now without explanation simply to provide pointers for the reader and to avoid gross misunderstandings.

I have just used such terms as "sociality" and "social order" a number of times, and it will help orient the reader if I offer a preliminary and rough specification of what is meant by them. By social, to begin with, I mean pertaining to human coexistence. Although the Latin *socialis* connoted companionship, the expression social has been used in a wider sense in modern times to qualify any

mode or aspect of human coexistence whatsoever. This broader usage is reflected in other definitions of social, such as one offered by
Parsons: "The 'social' is that element of the total concrete reality of
human action in society which is attributable to the fact of association
in collective life."[19] Human association, or interaction, is wider than
companionship and is thus a more adequate paraphrase of social.
Parsons's definition is nonetheless problematic, not only because he
defines social by reference to society, but also because, as we shall see,
not all features of sociality are attributable to the fact of human interaction.

Human coexistence, moreover, is people forming what is best described with the German word *Zusammenhang*. A *Zusammenhang* is a
state of held-togetherness. As suggested by the two words that render
the German expression in English, "nexus" and "context," a *Zusammenhang* is a hanging-together of entities that forms a context for
each. Human coexistence is a hanging-together of human lives that
forms a context in which each proceeds individually. This formulation is designed to accommodate states of sociality of varying
breadths and complexity. Lives hang together in the microsituations
of intimate relations, club activities, and classroom teaching as well
as within the wider macrophenomena of economic systems, artistic
practices, global communications networks, and international football. That by virtue of which lives in such formations hang together,
moreover, is obviously varied and complex.

It is important not to think automatically of a *Zusammenhang* simply
as interrelated individuals. All human nexus-contexts embrace individuals along with relations among them. It is an open question,
however, whether the hanging-together of lives is a matter of their
interrelations alone. While individualists affirm this thesis, wholists
counter with the claim that lives hang together also because of properties of the wholes in which they are enveloped. The position developed here contends that practices are the medium in which lives
interrelate. As this medium, they themselves are not simply interrelations among lives. Practices, consequently, are a dimension of human
coexistence distinct – though not separate – from individuals and
their interrelations. A *Zusammenhang* of lives is not interrelated individuals *simpliciter*, but individuals interrelated within and through
practices.

The term sociality is used in social thought in two principal ways.
It refers, first, to "socialness," the condition of being social; and in
this usage is opposed, for instance, to "individualness," or that which
pertains solely to the life of an individual. The sociality of human
existence in this sense is the social dimension or aspect of human
existence. The term refers, second, to the nature of the social dimen-

sion or aspect of human existence. Given my definition of social, sociality in this second sense designates the context-forming hanging-togetherness that constitutes human coexistence. Participation in such hanging-togetherness is thus what it is for a person to exist in a condition of sociality in the first sense.

The term social order, finally, refers to orderings in social life. By an "ordering" I mean an arrangement of entities in which each has meaning and place. A social ordering is an arrangement of human lives and of the things with which people deal in which people and things possess these properties. A kinship system contains such an ordering, as does an economic system, a baseball game, and also a momentary meeting in the street. Each of these social formations comprises an arrangement of people in which they perform inter-locking actions, are entangled in particular relations, and possess specific identities.

All social existence entails social order. Any context-constituting hanging-together of lives implicates an arrangement of those lives in which they are positioned with respect to one another. Arrangements of lives are thus built into *Zusammenhänge*. In coexisting people are ipso facto ordered. And whatever it is by virtue of which lives hang together and form nexus-contexts is at once the determinant of social order. The opposite of order in this sense is isolation, disconnect-edness, and, at the limit, chaos and randomness.[20]

This notion of social order must obviously be differentiated from the much more common notion of people living with one another successfully, compatibly, and without overt violent conflict. The op-posite of this more typical notion of social order is disharmony, breakdown, violent conflict, and what Durkheim called anomie. Not all cases of order in the first sense exemplify this more familiar notion. Order in the first sense, however, will be the sole object of concern in this book.

Many theorists have analyzed social order (and sociality too) along one or both of two axes. The first is normative regulation. In this vein, the theorist who in modern social thought most famously posed the "question of order" defined one of the two sorts of order he espied in social life as the social "process tak[ing] place in conformity with the paths laid down in the normative system" (where "norma-tivity" refers to ends, rules, and norms).[21] Theorists have also ana-lyzed order and coexistence by reference to the cooperation, rational-ity, and knowledge of individuals.[22] The prominence of these two axes reflects the ontological distinction between whole and individual. While individualism naturally avails itself of rationality and coopera-tion as media of sociality, many wholisms have adverted to norma-tivity in their accounts of coexistence (e.g., Hegel's *Sittlichkeit*). Conspicuously missing from both axes are understanding and the articulation of intelligibility. People's lives hang together not only through cooperation and rationality as well as conformity to ends, norms, and rules, but also through understanding and intelligibility.[23] Think, for instance, of how people brought up in the same culture typically understand objects in settings alike. No matter how ubiquitous and decisive normative regulation and cooperation-rationality-knowledge may be for human coexistence, this additional medium of coexistence accompanies and/or underlies their contribution to sociality and social order.

The discussion of sociality later in this book focuses on understanding and intelligibility and explores their omnipresent contribution to sociality and social order. This discussion, accordingly, makes no claims to be a complete theory of sociality or social order. (I use the term "theory" in the sense of a systematic and detailed account of some phenomenon or domain of phenomena.) The states of affairs through which people's lives hang together are legion. As noted, moreover, one lesson of recent thought is the appropriateness of skepticism in the face of general theories that claim to capture everything about something, at least where the something has any breadth or depth. My later remarks concern only certain fundamental features of sociality and social order. Indeed, far from representing a complete rendering of social life in general, my account of sociality, in specifying basic features of human coexistence, instead constitutes a framework through which to investigate social domains and phenomena and to uncover their further details and complexities. Similarly, the intervening account of mind/action and its constitution is not intended to describe everything about the rich diversity of conditions that are mentality/activity. Its ultimate aim is to analyze mentality only to the extent needed in order to appreciate its constitution through practices.

To conclude this introductory, contextualizing discussion, I want by way of an example to suggest how the account to be developed identifies aspects of sociality and mentality/activity that are overlooked or forgotten in contemporary social theories of various stripes. As quoted earlier, Zygmunt Bauman claims that a sociological theory adequate to the changed social constellations called "postmodern society" must resolutely refuse the concept of totality. The needed theory, he suggests, must instead approach contemporary life through the notion of agency within a habitat. In Bauman's portrayal, the postmodern world is composed of an immense plurality of agencies, each of which pursues ends and chooses means as moments in a lifelong course of self-constitution. By "self-constitution" he means a process of "self-assembly," in which agents consciously construct who they are (their identities) by cultivating their bodies and gathering "symbolic tokens" that denote membership in or fealty to various groups, movements, and ways of thinking. The pursuit of self-constitution transpires, moreover, within habitats that (1) are populated by myriad other agencies, none of which exerts overall managing or coordinating capacities, and (2) determine agency only in the sense that they provide resources, offer ends and means, contain reference points for symbolic tokens, and thereby set the agenda for the "business of life." The social world, or state of "sociality," that emerges out of the activities of these agents has, however, no "supra- or pre-agentic foundations."[24] The activities of the semiautonomous agents are alone responsible for whatever contingent and processual states of sociality exist at any point in time. The convergence with utilitarianism of this extreme version of postmodern nominalism is accentuated when Bauman writes:

Pluralism of authority . . . rules out the setting of binding norms each agency must . . . obey. . . . "Non-contractual bases of contract," devoid of institutional power support, are thereby considerably weakened. If unmotivated by the limits of the agency's own resources, any constraint upon the agency's action has to be negotiated afresh. (p. 201)

This quotation concerns the capacity of rules, understood as explicit statements, to govern action. But this very focus, together with the fact that Bauman starts from already fashioned and self-conscious agents, means that forms of action-limiting and -constituting "socialities" other than rules, available resources, and others' actions are simply left out of his theoretical sketch. Likewise, his comparison of social life, when seen from a bird's-eye point of view, with Brownian movement (p. 189) neglects how the practices discussed by practice theorists establish a sociality that always interconnects, constrains, and enables the "particles" of social life throughout their motion. Practices, to extend his analogy, institute "fields" that ipso facto govern these particles (agencies) throughout their motion. What's more, the only aspect of the constitution of agencies that appears in this sketch is the conscious attempts of agents to construct their identities. Again, this passes over the constitution of the minds and actions of agents that has always already occurred within practices out of agents' control – and often unbeknownst to them – whenever they set themselves to self-construction. Practices pervade and underlie *both* agency and habitat, weaving together agents and settings through the understandings and articulated intelligibilities that organize settings and govern activities, including those Bauman describes. Rejection of totality must not lead to a randomizing nominalism that, in isolating and disconnecting particulars, resurrects the self-contained unity from which the rejection fled in the first place.[25] Underlying and thus connecting all the particulars populating the social world is a web of understanding and intelligibility that establishes meanings and possibilities and thereby institutes agents while coordinating them with their milieus.

The following chapter begins working toward this account of sociality by developing a Wittgensteinian conception of mind/action.[26] Before setting to this, I acknowledge that much of what I will write is not explicitly found in Wittgenstein's remarks. My account is an attempt at creative interpretation, which draws out from a thinker's writings a position that is not officially presented there. I make no apologies for this procedure. It is valuable not only to summarize and analyze what a thinker says on given topics, but also to question what is presupposed and implied by these statements and, more radically, to coalesce and extend these thoughts in forms and directions unthought of by the thinker and perhaps even subject to his or her disapprobation. Indeed, to observe this procedure is to exhibit the highest respect for thought. Those critical of this modus operandi with regard to Wittgenstein's texts should recall Wittgenstein's remark in the preface to the *Philosophical Investigations:* "I should not like my writing to spare other people the trouble of thinking. But, if possible, to stimulate someone to thoughts of his own." Thoughts, in the current instance, that are nonetheless rooted in and inspired by his writings.

2

Mind/Action/Body

This work seeks to further comprehension of the nature of social life.[1] Why, then, begin with an analysis of mind/action? Since action is generally recognized to be a – if not the – central category in social ontology, every social theory operates with an understanding or explicit account of it and its "determination." What's more, the conceptions of determination figuring in these accounts invariably embody an interpretation of the mental. Even in the absence of an explicit theory of mentality, the very use of such concepts as goal, belief, emotion, desire, motivation, reason, scheme, expectation, disposition, habit, need, and consciousness evinces at least an implicit understanding of it. Of course, there are divergent interpretations of these concepts and assorted theories to inform their use. The multitude of conceptions of the mental operative in ontological thought is revealed by widespread disagreement over which subset of concepts represents the psychological context of behavior best.

It is not the want of social theorists to dwell on and fashion reflectively the understandings of mind/action on which their accounts of sociality rest. They instead typically import concepts and understandings from psychology (e.g., behaviorism and psychoanalysis) or, more frequently, everyday life. Everyday life traffic with common locutions for mentality and activity harbors an extensive repertoire of concepts and understandings of psychological conditions. More concerned with sociality than with mentality per se, social theorists smoothly draw on the concepts and understandings they acquired in mastering a language and growing up in a culture. Whereas some theorists – like the rest of us most of our lives – wield these concepts/understandings more or less u.•:eflectively, others transform subsets into systematic frameworks through which to examine human activity. An example of the latter procedure is Max Weber's typology of social action at the beginning of *Economy and Society*, which is constructed with such everyday concepts as rationality, purpose, and value. Many social scientists have also of course investigated the social determination of the phenomena designated by particular mind/action locutions, for example, emotion, reason, and belief. But although it would be cavalier to suggest that social ontologists do not ponder the nature of mind/action, too many simply borrow some framework

without considering exactly what that framework implies and assumes about human life, activity, mentality – and sociality.

So, one reason to commence with mind/action is to reassert the
central role that understandings of it play in social ontology. One
possible gain therefrom, in addition to enhanced self-understanding,
is increased judiciousness and forethought in the adoption of conceptual frameworks. There is, however, a second and for this study more
decisive reason to begin with mind/action. All social ontologies, and
not only individualisms, operate with a conception of the individual.
At the core of all such conceptions, I claim, lies the possession of
mind and the performance of action. Being an individual is above all
having mind and performing action, whatever else it might (necessarily) involve. Now, a variety of thinkers have maintained that mind is
socially constituted. If they are right, then the existence of individuals
is also a social matter. For the constitution of mind/action is *at once*
the institution of individuals. Consequently, fuller understanding of
the social nature of mind/action will yield deeper insight into the
social character of the individual subject, the phenomenon from
which all individualisms set out and to which all wholisms must at
some point descend. Comprehending the social determination of
mind/action will also afford access to the social ontological centrality
of practices. Conversely, I might add anticipatorily, mind, it will turn
out, is a "medium" through which practices are organized. This
result marks a further reason *ex ante* to begin with mind/action.

Several notions of what sort of thing one is talking about in speaking of "mind" are alive in the contemporary scientific context. In the
following, mind will be formally defined as what overall is attributed
to entities by the application of common mental locutions to them.
There is not, of course, a well-delimited set of such terms. Neither
the limits of "the mental" nor the outskirts of "the common" are
precisely determined. Many terms employed in psychological and
behavioral theories, for instance, are borrowed from the common
stock; and invented terminology or borrowed terms whose original
home did not concern individual human psyches (e.g., "repression")
migrate into common mental parlance. Despite the fact that the
common and scientific repertories overlap in ebbs and flows, an
extensive core of terms (e.g., "think," "want," "hope," "see," "hear,"
"desire," "expect," "believe," and "imagine") not only belong unproblematically to the category "mental expression," but also remain common regardless of how much science colonizes them. I emphasize
that characterizing mind as what overall is attributed to entities with
such locutions does not proscribe psychologists (and others) from
analyzing what mental matters consist in in either vocabularies of
their invention or common locutions transformed into technical

terms. It implies only that, so long as they seek to investigate mentality, their technical vocabularies and theories must concern that realm of phenomena in which the sort(s) of states of affairs picked out by the use of common locutions consist. (This manner of putting things will be clarified in the upcoming section on conditions of life.) The point of the preceding formal definition is not to exclude theoretical developments ex cathedra, but simply to give a preliminary identificatory indication of what domain of states of affairs is meant by "mind."

Some contemporary theorists of mind will still object that this demarcation is unwarranted handcuffing. Might some aspects of mind go unregistered in common mental locutions? How do we know that science won't substitute a theoretical vocabulary for our current common one?[2] These are large issues that cannot be examined here. In response to the first question, I repeat that the preceding formulation brooks theoretical vocabularies and only delimits the domain of phenomena that the theories involved should concern if they are to count as theories of the mental. In response to the second question, let me voice the conviction that so long as we *do* continue to use the common locutions, we will label them "mental," as we do today. That is, so long as people continue to employ these locutions, they will consider what they are talking about in doing so to be mind. If science causes us to jettison this vocabulary, this will show not that there is a better vocabulary with which to discuss mind, but that there is no mind.

There is, moreover, a further reason for the preceding definition that is telling for anyone concerned with the nature of social existence. Through the ages, not only philosophers, but essayists, writers, moralists, and social thinkers generally, have employed common mental locutions in discussing human life. It is a safe bet that they have always assumed that mind *is* what they were talking about in employing these terms. This remains largely true today in both philosophy and social theory. Again, this does not imply that philosophers and others have not also invented new terms for describing mind. But it does mean that the preceding formal characterization of mind is simply the going policy and position of the broad tradition of Western philosophy and social thought.

Overview

Traditional philosophies conceived of mind as the ontological substrate or site for the functions and attributes called "mental." Examples of this substrate or site are *psuchē*, the soul, thinking substance, and consciousness. Each of these constituted not only a home for mentality distinct from objects in the world, but a principle for relating and unifying the panoply of mental acts and states. The sphere of mentality was bounded and discrete, and it exhibited the kind of structured coherence associated with real entities or locations.

One prominent moment in twentieth-century thought has been an assault on unified ontological sites for the psychological, on the idea that mind is a substance, place, or realm that houses a particular range of activities and attributes. Although Wittgenstein played a key role in this critical movement, his ideas about the correct conception of psychological phenomena are equally significant. He does not tread the standard paths of denying their existence, reducing them to behavior, identifying them with states of the body, treating them as explanatory functional or computational states, or conceiving them as theoretical entities. He instead views them as conditions of life expressed by the human body.

On my reading, Wittgenstein's texts suggest that what is accomplished in using common mental locutions is the articulation of how things stand and are going for people in their existence, in their continuous moment-to-moment involvement with persons, objects, and situations. Mind, consequently, is how things stand and are going for someone; and mental phenomena (e.g., believing, hoping, expecting, and seeing) are aspects or ways of this. I will refer to mental phenomena as "conditions of life." The term "condition" is used here in the sense of the state of something's being, its "how-it-is." It is not used in the sense of a prerequisite for something's appearance or occurrence (as in Immanuel Kant's claim that the concept of an enduring object is a condition of knowledge of the world). A condition of life is life being thus; and the being thus in question is things standing and going someway. In characterizing mental matters as life conditions, consequently, I am not claiming that they are indispensable for life. I am maintaining only that they are aspects and ways of a particular dimension of human existence: how things stand and are going for someone.

These aspects and ways, moreover, are expressed by bodily doings and sayings. To say that they are expressed in bodily activities is to say that these activities make them present in the world. Wittgenstein sometimes puts this as follows: In specific circumstances, a particular condition of life *consists in* particular bodily activities. I emphasize that, in the following, the term "bodily activities" connotes not simply bodily doings and sayings, but also what Wittgenstein calls "fine shades" (*feine Abschattungen*) of behavior, such as the manners in which doings are carried out and the tones of voice with which something is said.

A further feature of Wittgenstein's account, to be examined in detail in the subsequent chapter, is the social constitution of mind. Which aspects of how things stand and are going are expressed by particular bodily activities depends ultimately on the social practices in which people learn to perform these activities and to take interest in and react to others' performances. Connections and orders among mental conditions, consequently, are laid down in practices. The "structure" of mind is established in these practices and does not derive from the intrinsic natures of the substances, realms, attributes, and inhabitants countenanced in traditional theories. Although some mental phenomena are biological in origin, and although the body is their medium of expression, they and their interrelations and patterning are for the most part socially instituted.

That mind is social is by now a familiar idea. Many contemporary theories view mental states or cognitive operations and processes as in some sense socially constituted. Profound differences, however, separate Wittgenstein's account from these theories. For instance, most contemporary theories treat mind as either an abstract apparatus or an underlying cognitive one that is functionally or causally responsible for bodily activities. This apparatus no longer possesses the substantial being envisioned in the work of Descartes and his followers, but it retains a unity defined by relations both among its components and between these components and bodily activities and processes. In Wittgenstein, by contrast, mind is a collection of ways things stand and are going that are *expressed* by bodily doings and saying. Not only does this collection carry no presumption of unity, but the expressive relation is not causal: Conditions of life do not cause the activities expressing them. The activities are instead that through which how things stand and are going is made present in the world. So an activity does not express a given condition by virtue of being its effect. It does so instead, as we shall see, by virtue of its place in the weave of behaviors and of occasions and contexts of behavior.

A second difference concerns common mental locutions. For many contemporary theorists espousing an apparatus conception of mind, ordinary language mental idioms designate states of the apparatus that are responsible for bodily activity. "Mental state" talk is then construed as pseudoscientific discourse that embodies a nascent theory of human beings and explains bodily doings and sayings by reference to discrete entities. For Wittgenstein, by contrast, such words as "pain," "joy," "belief," "doubt," "thinking," and "hope" are used to articulate how things stand and are going for someone. The conditions thereby articulated are states of affairs, not entities; so the use of these terms does not pick out states of an abstract or real, underlying apparatus. As indicated, moreover, these states of affairs are expressed in bodily activities. Indeed, the use of these words is keyed to behavioral expressions and patterns thereof – using them articulates how things stand or are going with people who exhibit certain behaviors. For Wittgenstein, in fact, behavior and expressions on the one hand and understanding and language on the other are bound together within social practices. Aside from biologically based naturally expressive behaviors, the performance of behavior that expresses such and such a condition presupposes practices on the background of which others are able, on the basis of that behavior, to understand and say that this is the actor's condition; and people are able to understand and say this by virtue of participating in these presupposed practices. So there is nothing theoretical about "mental state" talk, and its explanatory powers are not tied to the operations of an underlying apparatus. Instead, such talk is the shared, original, medium of the everyday explicit understanding a person has of herself and others.

For Wittgenstein, "mind" is how things stand and are going for someone. These matters are expressed in bodily activities and formulated in "mental state" language. Mind is thus constituted within and carried by practices, where these locutions are used and people acquire the abilities and readinesses to perform and to understand a range of bodily doings and sayings. As something essentially so expressed and articulated, mind is primarily instituted, and achieves any degree of complexity, only within social practices.

The body obviously plays a key role in Wittgenstein's analysis of mind. This fact once led some interpreters to treat him as a reductive behaviorist à la Gilbert Ryle, a view from which Wittgenstein explicitly and repeatedly distanced himself. As most commentators recognize today, Wittgenstein is neither a behaviorist, claiming that mind is nothing but bodily activity,[3] nor a mentalist, maintaining that mind is a distinct and substantial substance or realm. On my interpretation, the body is an entity that, in its doings, sayings, and sensations, manifests and signifies psychological states of affairs. This position is neither behaviorist nor mentalist, since aspects of how things stand and are going for someone are neither identical with behavior nor components or states of a distinct substance, realm, or apparatus.[4] They lack the discreteness, structure, and/or substantiality required to be identified as such types of entity. Wittgenstein does agree with reductive behaviorism against mentalism that there is only one realm, not two, of entities or processes involved, that of bodily activity. Unlike reductive behaviorists, however, he denies that mental conditions are nothing but such activities. Mental conditions are *expressed* by these activities. As an underlying biological, but socially molded, work, the human body is the site for the expression of the aspects of human existence articulated in "mental state" talk. This position places Wittgenstein in the company of such theorists as Maurice Merleau-Ponty and Helmut Plessner, with whom he is not standardly associated and whom he most probably never read.

Wittgenstein's explorations of mentality offer a fresh perspective on the ontological status and structure of mind, its relation to the body, and its social constitution. In addition, they provide new insight into the distinction between the inner and the outer and expose the artificiality of the distinction between the realms of mind and action. Traditional dichotomizing described the mind as the "inside" of a human being, while designating behavior as his or her "outside." Wittgenstein famously disputes the Cartesian-Christian sense of an inner realm, although, as we shall see, he acknowledges the existence of inner phenomena. He invigorates, moreover, the notion of the inside of a person's *life*. This inside is not a realm or entity, but how things stand and are going in life. One person is more attuned than a second to the inside of the second person's life when he has a better understanding than she of what she is about. The fact that in most situations people have the best and surest understandings of what they are about promotes the illusion that there is a private inside realm of mental entities accessible to their possessor alone.

Finally, whereas the idea that mind is expressed by bodily happenings might be unfamiliar, the idea that bodily doings constitute actions is commonplace in contemporary theories of action. In Wittgenstein, as we will see, mental states, intellectual attitudes, *and* actions are all aspects of how things stand and are going for someone and expressed in the same ways by bodily doings and sayings. This parity, in conjunction with the dissolution of the inside/outside dichotomy traditionally associated with mind/action, makes evident the artificiality of any categorical distinction between mind and action.

Conditions of Life

Wittgenstein abjured metaphysics, yet his ideas on mind reflect a change in ontological proclivity. Throughout the history of Western philosophy, the mind, or soul, was treated as a thing. The precise nature of this thing was, of course, problematic. Yet it shared with all things the quality Heidegger called *Vorhandenheit*, abidingness. The mind abided by virtue of the continuing existence of essential features. These features might be a faculty framework (e.g., reason, passion, spirit), a will, an I (that thinks), or the stage upon which ideas come and go. In each case, the mind, as abidingly there, constituted a

site or home for the phalanx of such fleeting or slowly evolving
mental attributes as desires, hopes, pains, thoughts, moods, and personality.

Nonsubstance ontologies existed during the modern era, but they
were mostly ignored in the context of theorizing mind. Indeed, the
very expression "the mind" solidifies the dominance of substance
ontology. By the late nineteenth century, however, a number of
thinkers had adopted a more process-based approach to mentality.
Prominent among these were Friedrich Nietzsche, Wilhelm Dilthey,
and Henri Bergson. No longer an abiding self-identical persister,
mind was now understood as, or as a dimension of, a temporally
extended, continuously developing process whose identity lay not in
the persistence of essential features but in the sinews of continuity
linking its phases. Mental phenomena, accordingly, were no longer
viewed as attributes of an abiding substance, but as episodes in an
unfolding process. Constituting the being of which they were phases,
they no longer were features that could come and go as their bearer
remained the same. This picture claimed manifold adherents during
the twentieth century, prominent among them John Dewey, the life-
philosophers, Wilfrid Sellars, and Ludwig Wittgenstein.

In suggesting this of Wittgenstein, I do not maintain that he acquired this picture from any of these thinkers, not even that he read
any of them (though we know the contrary). I aver, instead, that
from the vantage point afforded by historical distance, his ideas can
be seen as swimming in the emergent stream of process-metaphysical
approaches to mind. For Wittgenstein, mind is particular dimensions
or aspects of the process of human life. "I would like to say: Psychology deals with certain *aspects* of human life" (*RPP* II, 35). As indicated, the dimension of this process that is mind is how things stand
and are going for someone, which is expressed in bodily doings and
sayings and spoken and written in common mental locutions.

Admittedly, Wittgenstein nowhere explicitly declares life the process of which mentality is certain aspects or dimensions. And nothing
in what follows rests on the veracity of this speculative interpolation.
But the prominence of flow imagery in his later work, together with
the repeated references to life and continuing importance of Arthur
Schopenhauer for all his thinking,[5] suggests that some sort of life-
process framework forms the context for his particular approach to
and observations concerning mentality. In Wittgenstein's texts, the
image of a stream of life either is explicitly invoked to name the
general, constitutive context in which words and gestures have meaning and phenomena such as meaning are what they are,[6] or implicitly
operates as a model for understanding mentality. The former point

is familiar. The latter is hinted at in passages that draw an analogy between mental phenomena and states of flowing water:

Take the various psychological phenomena: thinking, pain, anger, joy, wish, fear, intention, memory etc., – and compare the behavior corresponding to each. . . . Isn't this as if someone were to say: "Compare different states of water" – and by that he means its temperature, the speed with which it is flowing, its color, etc.? (*RPP* I, 129–130)

The occasional appearance in his texts of what one might call an adverbial theory of mind reinforces the impression of a process orientation:

Comparison of bodily processes and states, like digestion, breathing etc. with mental ones, like thinking, feeling, wanting, etc. What I want to stress is precisely the incomparability. Rather, I should like to say, the comparable bodily states would be *quickness* of breath, *irregularity* of heart-beat, *soundness* of digestion and the like. And of course all these things could be said to characterize the behavior of the body. (*RPP* I, 661; cf. 284, 275, and *LW* II, p. 21b)

Mentality is here compared with features of the occurrence of behavior and bodily processes, as opposed to the behavior and processes themselves. Wittgenstein is not, of course, reducing mind to aspects of the flow of human behavior and bodily processes. Rather, he is saying that if one wants to find an analogy for mind in the realm of the body, then the proper analogue is adverbial features of these occurrences. This raises the question, Of what process is mind adverbial aspects? Life, it appears to me, is the obvious candidate.

As a debunker of the picture of an inner realm or space called the mind, Wittgenstein emphasizes the public, in-the-world character of the stream of life. In this way, he differs from thinkers such as Dilthey and Bergson who conceived the flow of life primarily as a stream of consciousness, a continuous succession of linked states of consciousness. For Wittgenstein, life is a continuous flow of bodily activity in the world directly accessible to persons other than their performer. It turns out, however, that life has, so to speak, two faces: a continuous behavioral one "open to view" in the public world, and an intermittent "inner" one accessible only to its possessor. To understand this, we need to consider Wittgenstein's notion of appearance.

Although he nowhere thematizes his usage, Wittgenstein's repeated use of the terms *Erscheinungen* and *Phänomene* (usually translated as "phenomena" or "manifestations," occasionally as "appearances") is redolent of past usages in modern German philosophy. With these terms, he appears to mean experiential events and states of affairs. As he at one point pseudodefinitionally remarks, a phenomenon is something that can be observed (*RPP* II, 75; cf. *LW* II, p. 76c). The phenomena of X are thus the experiential, spatio-temporal or temporal presences of X. In this vein, he speaks of the phenomena of language, electricity, and experimentation, and of psychological phenomena.

Another word for phenomenon is appearance. Mental phenomena are appearances of mind. These appearances, however, are not experiential occurrences that announce something that also has a being (in-itself) not shown in experience. Wittgenstein rejects this Kantian conception of appearance in favor of an alternative notion found in J. W. von Goethe, that appearances themselves constitute reality. There is nothing behind the appearances encountered in experience. Appearances are not simply how something manifests itself to us, at the same time "holding back" something of itself. Experience is the arena in which reality shows itself as what in itself it is.[7]

According to the mediative theories of experience dominant in the mid to late twentieth century, all experiential encountering of objects is mediated by concepts or understanding. In his very late thoughts on perception, by contrast, Wittgenstein outlines elements of a "realist" viewpoint that maintains that reality is directly ("bodily," as Husserl put it) encountered in perception. Conceptual and interpretive mediation infect instead the *descriptions* people offer of what is perceived. A seen object, for example, is subject to multiple descriptions, each of whose specification of "what is seen" is mediated by interests, concepts, and ways of understanding. These specifications, however, contrast with the bodily seen, mute reality that cannot be put into words (cf. Wittgenstein's references to the aroma of coffee and the taste of sugar, e.g., *PI*, 610; *Z*, 654). Unfortunately, it would take us astray to consider further this perspective on perception.

Two orders of life phenomena appear in Wittgenstein's remarks. One is a realm of "outer" phenomena, there in the world for others to see. The second is a realm of "inner" phenomena, accessible to their possessor alone. The appearances of human life thus include gesticulations, speech acts, facial expressions, and tone of voice, on the one hand, and sensations, feelings, images, and words "before the mind," on the other.

It is important to emphasize that the realm of inner phenomena is not a sphere of "mental" entities as conceived in the Cartesian era. To start with, note the limited range of types of episodes that are inner phenomena. In contrast to arch-Cartesians such as Locke and Hume, Wittgenstein does not treat emotional conditions such as joy, fear, and despair or cognitive attitudes such as doubt, belief, and certainty as inner episodes. Their relation to this realm will be examined shortly. Moreover, the *episodes* that are inner phenomena, episodes such as tickles, kinaesthetic feelings, sense sensations, and images, must not be assimilated to *objects* encountered in the world, for instance, rocks, stairs, bodies, and telephones. The difference comes out in the ways entities of these two categories are identified.

Objects encountered in the world can be and often are identified on the basis of their features. One can examine such objects in perception and ponder what they are on the basis of perceived and inferred attributes. Nothing like this can take place when identifying inner phenomena. Wittgenstein's private language argument demonstrates that alleged acts of inspecting, inwardly pointing to, and recognizing inner phenomena are at worst impossible and at best irrelevant to the employment of the terms designating these phenomena. One should not, therefore, speak of identifying inner phenomena on the basis of directly apprehending their features. In place of an inner inspection model of self-knowledge, Wittgenstein maintains that the cognizance and self-attribution of pains, images, seeings, and the like are, in the usual case, criterionless. They are based on nothing at all. "Knowing" that one is in pain, consequently, is not a matter of recognizing (features of) particular inner episodes. One just "knows" (or cannot doubt) that one is in pain; and, in fact, it is because one "knows" this that one "knows" (or cannot doubt) that the currently occurring inner episode is pain. This point about self-knowledge is repeated vis-à-vis colors and images. One does not identify a color as red and an image as one of Pete by apprehending and recognizing features of the color or image and drawing conclusions. Rather, one just "knows" (or cannot doubt) that one is seeing red or imagining Pete and, as a result, can identify the current color as red and the current image as of Pete (see *Z*, 32, 481, 483, 498, 663; *PI*, p. 185; *LW* II, p. 5a–b). Inner phenomena such as pains, images, and sense impressions, consequently, are identified only "derivatively" on the basis of a person's "knowledge" (or inability to doubt) that he is in pain, imagining Pete, seeing red, and so on. Even this formulation, however, understates Wittgenstein's point. For it is not the case that there is some inner episode on the one hand and knowledge of one's condition on the other, and that the identity of the former is ascertained through the latter. Rather, there *is* an inner episode at all only insofar as one "knows" (or cannot doubt) that one is having a sensation or sense impression. This means that there being pain or an impression of red (or even just "something") is a matter of one's "knowing" that one is in pain or seeing red (or experiencing something).[8] All this, incidentally, arises from the fact that a person has sensations and images only so long as she is aware of them (see, e.g., *PI*, 321; *RPP* I, 208).

The categorical difference between inner episodes and objects in the world is also revealed by the fact that inner episodes cannot be observed in the way objects can, namely, by attentively following alternations in their features (see *LW* I, 618). Inner episodes are observed only in the sense that they occur consciously.

In sum, Wittgenstein's criticisms of the distinction between the inner and the outer do not affect his acknowledgment of sensations, images, feelings, and the like. For his reproofs are of a Cartesian inner, a realm of inner objects that are identified on the basis of apprehended features and are capable in some sense of bringing about outward behavior. Inner episodes are not objects. Moreover, the relation between them and behavior, as we shall see, is not one of causality.

None of this, by the way, denies that inner episodes have a bodily encountered mute reality. It simply means that a person's relation to these episodes is not one of observation, recognition, and inspection. A person cannot adopt these relations to inner episodes because, in a sense, a person *is* these episodes. Consequently, there does not exist the sort of gap between experiencer and what is experienced that in the case of external objects is overcome through attention, exploration, and inspection. A less metaphorical way of putting this point is that inner episodes are the appearances of a person's conditions to himself. As appearances of himself to himself, a person's relation to them is consummated in their conscious occurrence.

Mentality, consequently, is particular aspects of the process of life, which itself appears as a continuous stream of behavior and an intermittent flow of sensations and images. I want now to give greater flesh to this conception through the notion of a condition of life. Wittgenstein liberally applies the word *Zustand* (translated as "state") to mentality. He designates pains, images, seeings, and the like as states of consciousness *(Bewusstseinszustände)*, and characterizes emotions and moods as well as forms of conviction such as doubt and belief as *Zustände*. "State," however, is an inadequate translation of a word that covers such disparate aspects of human life.[9] Interestingly, this translation would make sense in a substance-based ontology. Since all these different aspects of mental and intellectual life would then be viewed as attributes, functions, or occupants of a substance or realm, they could be conceived of as components, subconfigurations, or units ("states") of its construction, structure, or occupation. Wittgenstein attacks this picture of mind, which he sometimes labels the "apparatus" picture. To capture the ontological shift effected in construing mentality as aspects of human life, *Zustand* should instead be translated as "condition." "Condition" lacks the connotations of structural discreteness carried by "state" (though some uses of the latter, e.g., "the state of things," have the proper connotation). All

psychological states, thus being in pain, doubting, being joyful, believing, hoping, and so on, are conditions of life, or ways of being.
They are aspects of how things stand and are going for someone in
his existence rather than elements or structural configurations of a
substance or realm.

The largely nondiscrete, nonstructured, and nonexperiential nature of life conditions reflects the capacity of poetry and pictures to
depict them. Wittgenstein returns to this point on a number of occasions:

> It is important, however, that there are all these paraphrases! That one can
> describe care with the words: "Ewiges Düstere steigt herunter." I have per-
> haps never sufficiently stressed the importance of this paraphrasing.
> Joy is represented by a countenance bathed in light, by rays streaming
> from it. Naturally that does not mean that joy and light *resemble* one another;
> but joy – *it does not matter why* – is associated with light. (*RPP* I, 853; cf.
> 377, 1088)

Objects, apparatuses, and their structural configurations can be described in poetry and pictures. But these entities lack the dimension
of how things stand and are going that is characteristic of human life
and captured in poetic phrases and symbolic images. Since human
beings customarily use common mental vocabulary to speak and write
about the condition of their existence, the use of this vocabulary
more closely resembles the use of poetic phrases and symbols than
that of theoretical explanatory vocabulary.

To describe the connection between the conditions and appearances of life, Wittgenstein employs the terms *Ausdruck* and *Äusserung*
(expression). The outer and inner episodes comprising the stream of
life express conditions of life. To say that episodes express these
conditions is to say that they make these conditions present in the
world (the realm of phenomena). Nervous fidgeting, for example,
can make present anxiety, expectation, fear, or any number of ways
things are going. Conversely, conditions of life *consist in* (*bestehen in* –
a common Wittgenstein locution) particular inner and outer episodes. To say that a condition consists in such episodes is to say that
these episodes are what in certain circumstances there is in the world
to this condition. Fear, for example, might consist on a particular
occasion in grimacing, shaking, feelings of tension, and a gnawing in
the stomach, whereas believing in God might consist in a given situation in praying, genuflecting, and the occurrence of certain images.

Many concepts of life conditions, moreover, are coordinated with
patterns of life (*Lebensmustern*) that repeatedly emerge in the weave
(*Teppich*) of life (see, e.g., *PI*, p. 174). These patterns comprise behavior, utterances, and their occasions, and are completed by a picture
of the inner (see *RPP* II, 650–52). The reason why a picture of the

inner, and not the inner itself, completes the pattern is that any
pattern with which language use is coordinated must be publicly
accessible, and people's inner episodes are not available in the way
their behavior is. A picture of the inner nonetheless comes into play
since conditions of life are expressed not only in behavior, but also in
sensations and images. Life patterns are most commonly associated
with and most well defined vis-à-vis emotions, moods, and states of
consciousness. They are not common or definite in the case of cognitive conditions such as belief. It is important to note that, according
to Wittgenstein, not all patterns characterizing people's lives are captured in their concepts. Which extant patterns are so articulated
depends on people's interests, what they react to, what matters to
them, and generally, the traditions of behavior and attribution of
which they are the heirs (on this and the following points, see Z,
378ff.). The patterns most likely to be overlooked are those with
strong biological or physiological roots. Patterns that rely heavily
on social practices, in the sense that people come to perform their
constituent behaviors through social training and learning, almost
always have correlative concepts. Notice that the ranges of patterns,
life condition concepts, and possible conditions characteristic of a
group of people are for the most part products of past and current
social interaction among members of that group.

I should further note that the patterns corresponding to condition-
of-life concepts are invariably irregular and indefinite. This means
that the constellations of doings, sayings, and occasions to which
attributions of being in pain, being joyful, or believing X (etc.) respond vary tremendously. Further undermining the determinateness
of particular patterns is the close interwovenness of the patterns
associated with different condition-of-life concepts. This complexity
raises the question of how human communities manage to continue
using these concepts, an issue to which we will return.

Now, conditions of life are not identical with the life phenomena
expressing them. "The psychological verbs to see, to believe, to think,
to wish do not signify phenomena" (Z, 471). " 'But "joy" surely designates an inward thing.' " "No. 'Joy' designates nothing at all. Neither
something inner nor something outer" (Z, 487; translation modified).
Joy is, instead, *expressed* by outer and inner episodes, meaning that (1)
these episodes make joy present in the world, and (2) there isn't
anything more in the world to being joyful than these episodes.
Another way of putting this is that these episodes are the spatiotemporal and temporal appearances in the world of joyfulness. As noted,
however, there is no deeper reality of which these appearances are
emanations or representations. But this, in turn, does not imply that
being joyful is nothing.[10] Joyfulness is a way things are going, in

Heidegger's terms, a way of being(-in-the-world). As such a condition, it is made present in the world through behavior, sensations, and images. A condition of life is not identical with its appearances, but it is also not an element in the world in addition to them.

Rom Harré expresses a related thought when he writes that there is no such thing as "an emotion." What there is are various ways of acting and feeling emotionally.[11] To be, say, fearful is to have appropriate feelings and to perform appropriate behaviors, not to be characterized by or to have a certain property (fear). (For Harré, which feelings and behaviors are "appropriate" to a given emotion is established by a specific social role, the assumption of which is tied up with "having" that emotion.) This formulation captures the nonphenomenal status of life conditions. I do not know whether Harré would also join Wittgenstein in describing these feelings and behaviors either as the appearances/expressions of emotions or as the worldly phenomena in which emotions consist in particular circumstances.

It is important to emphasize that the relation of expression (manifestation, making present) is noncausal. The appearances of a life condition are not caused by that condition (cf. Z, 526; RPP II, 324). "And if the play of expression develops, then indeed I can say that a soul, something inner [i.e., conditions of life] is developing. But now the inner is no longer the cause of the expression" (LW I, 947). There is nothing behind the appearances for Wittgenstein. The life conditions of which phenomena are appearances are not an underlying reality that causes the appearances, but ways of being expressed by them. So phenomena do not come to be appearances of a given condition by being its effect. They instead acquire this status from their place in the play of phenomena, occasions, and existing life conditions, all this contextualized within practices of reaction, attribution, and education (cf. the impending discussion of the contexts constitutive of expression). To say, on the occurrence of a certain behavioral episode, that someone is in a particular condition is to make explicit something the behavior expresses about how things stand in his or her life, not to explain causally why it occurred. Incidentally, Wittgenstein believes that the appearances of life conditions have causes, to be discovered by the sciences of neurobiophysiology. What these sciences uncover, however, are not life conditions, for these conditions are socially instituted ways of being, not yet-to-be-discovered causally efficacious bodily states and pathways. The human sciences – as opposed to neurobiophysiology – are the fields specializing in the disclosure of life conditions.

If conditions of life are not spatiotemporal phenomena, and yet there is no reality behind such phenomena, what are they ontologically? A condition of life is an event: things standing or going some-way for someone. Bodily activities, consequently, are the worldly phenomena in which events of this sort consist. (The necessary over-lap and interconnectedness of such events over a period of objective time marks a manifold of them as a process.) Another way of saying this is that a condition of life is a state of affairs: that things stand or are going someway for someone. What bodily activity expresses, consequently, is certain types of state of affairs that involve people. This formulation makes manifest that life conditions are not realities standing behind bodily activities, whose being is not fully expressed in those activities. There cannot be anything to things standing or going someway for someone that goes unarticulated in the behavior expressing it. For, when expressed, this state of affairs shows itself as what in itself it is, namely, a particular state of affairs, things going or standing someway for someone. Unexpressed in the bodily activities concerned can be at most *further* states of affairs related to the ex-pressed one. So a condition of life is neither a phenomenon nor a reality behind the phenomena. It is a state of affairs that, in particu-lar circumstances, consists in, is expressed by, particular bodily activ-ities.

This account of life conditions underwrites a definition of per-sonhood. Parallel to Strawson's analysis,[12] a person can be defined as the type of entity to which both life conditions and physical proper-ties can be (correctly) ascribed. (That this formulation is not alien to Wittgenstein's thinking is shown by *PI*, 573; cf. 286.) By physical properties I principally mean physical states of a person's body such as weight, red blood cell count, renal functioning, and so on. This definition, as it stands, is incomplete, since it admits many sorts of animals as people. A fuller definition would specify further charac-teristics that an entity must bear to qualify as a person (e.g., identity and gender, the grasp of language, and/or self-consciousness). Re-gardless, however, of whatever further characteristics are argued to be definitive of personhood, conditions of life remain the central component of this status, for possession of all putative additional defining characteristics presupposes them. Nothing is self-conscious, gendered, or a master of language unless it performs actions and is in mental and intellectual conditions. So, because life conditions are a necessary condition of all other essential traits of personhood,[13] being a person is *above all* being the type of entity to which life conditions and physical properties can be ascribed. Notice that since conditions of life consist in behavior and inner episodes, a person is an entity to which certain bodily activities, bodily sensations and feelings, and bodily states can be ascribed. This formulation empha-sizes the role of the body as the seat of mentality (and activity, too, as we shall see). Notice, further, that this analysis does not rest on metaphysical presumptions or positions. It simply records the fact that the creatures we call people – that is, of whom we say that they act and have minds – are creatures like us, entities who are in-the-world via behaving and feeling bodies (cf. *PI*, 283–284).

A person is not a substance or inner kernel. As will be clarified later, individuality is a socially constructed and achieved status. Personhood is an *effect* of social practices, in that expressive bodies, life conditions, and ascriptions/comprehension of these conditions exist (for the most part) only within practices. The above analysis entails, further, that an individual's "coherence" or lack thereof is staged entirely in the play of his or her life conditions and bodily states. As a result, identity, in the sense of continuity and self-identity, is a tendentious affair. There can be no presumption of unity in the play of conditions and states, and whatever coherences are sustained there go unanchored in an abiding, substantial self. In this regard, the preceding analysis of personhood converges with Judith Butler's performance theory of gender identity. She writes:

> Whereas the question of what constitutes "personal identity" within philosophical accounts almost always centers on the question of what internal feature of the person establishes the continuity or self-identity of the person over time, the question here will be: To what extent do *regulatory practices* of gender formation and division constitute identity, the internal coherence of the subject, indeed, the self-identical status of the person? . . . The appearance of an abiding substance or gendered self, what the psychiatrist Robert Stoller refers to as a "gender core," is thus produced by the regulation of attributes along culturally established lines of coherence. . . . But if these substances are nothing other than the coherences contingently created through the regulation of attributes, it would seen that the ontology of substances itself is not only an artificial effect, but essentially superfluous.[14]

I will return to this parallel later.

I indicated that behavior, sensations, and images express particular conditions of life in specific circumstances. Which condition(s) a given phenomenon expresses always depends on the context in which it occurs. An obvious example is tears, which express different emotions in different circumstances. Generally speaking, the expressive relation is constituted by contexts of four types. The first is the past and future behavior (and inner phenomena) of the person involved. For example, whether Mike's utterance "I hope he comes" expresses hope that John will come or hope that Jack will might depend on, for instance, whether he was just looking at a picture of John or Jack, or on whether he expresses joy or disgust on John's arrival.

The second type of context is the web of life conditions that already holds of the individual. At any point in time, a person is in a particular, though not precisely delimited, array of conditions. This wider

set helps determine which condition(s) are expressed by current behavior, for behavior can (usually) only express those conditions that
mesh with the conditions the individual is already in. (I write
"meshes" instead of something more definite such as "rationally co-
heres" since it overly narrows the possible combinations of life conditions to insist, à la Donald Davidson[15] for instance, that they uphold
principles of rationality.) For instance, suppose Caroll is seen berat-
ing one of his students for an inept term paper. Caroll, believing that
he is particularly clear-minded that day, might claim when asked that
his behavior expressed frustration with the student's bad performance. It might be, however, that Caroll just received a low merit
evaluation and feels somewhat dispirited as well as distraught and
that what his behavior actually expressed was frustration with his
own performance.

The third type of constitutive context is the immediate and wider
situations in which people act. That, for instance, pushing someone
out of the way at the airport expresses disdain instead of the desire
to wrestle might be partially determined by the fact that the accosted
person is a proselytizing religious fundamentalist and not, say, one's
wrestling teammate from college. Or, to build on an example of
Wittgenstein (*PI*, 581; *Z*, 67), that a person's nervous pacing at the
airport departure gate expresses fear of an explosion and not, say,
the desire to annoy those around him, might depend in part on his
having overheard two people whispering the day before, "Tomorrow
at ten o'clock the fuse will be lit."

The final constitutive context is the practices in which people participate. Although this context will be extensively discussed in the
following two chapters, some preliminary and clarificatory remarks
are in order. Practices embrace behavior, speech acts, training, and
learning. Participating in them makes it possible for a person to be in
life conditions that do not have natural expression(s). The range of
possible conditions, consequently, that behavior can be understood to
express is delimited by the range of practices; and cultural variation
in practices underwrites variation in possible conditions. Social practices also set up expressive connections between specific behaviors
and conditions. Some behavior, for instance, is an extension, or aug-
mentation, of natural reactions, for example, utterances of "Ouch!"
or "That hurts!" which augment the crying of a baby in pain. That
these utterances express pain, and not joy or doubt, lies in how they
were grafted by cultural training upon the natural expression of
pain. Beyond this, a wider range of conditions, including many states
of consciousness and most emotions and moods, have characteristic
behavioral expressions and are associated with patterns of behavior,

utterance, and occasion. The institution of both characteristic expressions and life patterns is also largely the work of practices. The social background of understanding-carrying and intelligibility-articulating practices also envelops the first three contexts. For practices underlie, if not directly determine, understandable patterns of past, present, and future behavior, intelligible combinations of life conditions, and the relevance of the immediate and wider situation to expressivity. Within practices (1) intelligible or paradigmatic patterns of behavior, combinations of conditions, and situational relevancies are laid down and lived through; (2) people's behavior becomes informed by these patterns, combinations, and relevancies; and (3) people come to understand everchanging patterns, combinations, and lines of relevancy, as well as the conditions of life that bodily activity expresses on their basis.

Now, there are four main categories of life conditions.[16] Three of these have appeared in the preceding discussion. What differentiates all four categories from the physical states that can be attributed to an individual are ascriptional asymmetries: Whereas others ascribe them to a given individual on the basis of that individual's behavior, that individual (usually) ascribes them to herself on the basis of nothing at all (at least nothing from the perspective of traditional theories of empirical knowledge). The first category is conditions of consciousness (*Bewusstseinszustände*) such as being in pain, feeling itchy, imagining, seeing, and hearing. The second category is emotions and moods such as being joyful, being happy, fearing, depression, and anxiety. Wittgenstein names these first two categories "mental conditions" (*Seelenzustände*). What warrants this appellation is, in his view, their possession of "genuine duration" (e.g., Z, 78), which I interpret as uninterrupted expression throughout their existence. Because mental conditions possess this property, one can determine how long they last by means of a stopwatch. Conditions of consciousness consist in the continuous occurrence of some such phenomenon as a pain, image, or impression. Because of this, these conditions can be interrupted by a break in consciousness or a shift in attention; and although one can attend to their course, one does not have to convince oneself through spot checking (*Stichproben*) whether or not they persist (e.g., RPP II, 45, 57). Emotions and moods, meanwhile, are continuously expressed by a combination of behavior, bodily sensations, and images. A break in consciousness or shift in attention might not, consequently, interrupt them. Many of the behaviors and bodily sensations that express emotions, moreover, are characteristic of the particular emotions involved.

The third category of life conditions is cognitive, or intellectual (*geistige*),[17] conditions such as doubting, believing, being certain, wanting (*Wünschen*), intending, remembering, and understanding. Although such conditions exhibit considerable and significant differences among themselves, they are all attitudes and stances. They are not mental conditions. For they lack "genuine duration," that is, are not continuously expressed when a person is in them. They do endure, of course, but this does not involve continuous happening. It has more to do with continuities in what a person would do and say under certain circumstances. Furthermore, although some cognitive/intellectual conditions (e.g., doubt) have characteristic expressions, others (e.g., belief) do not. All, however, are expressed in behavior (and at times sensations and images), as evidenced by the fact that a person's life shows what he believes, doubts, and understands (e.g., *OC*, 7, 431). Note that some conditions of life (e.g., love, hate, expectation, hope, and even belief) fall at times into the second category and at other times into the third, depending on whether they are continuously expressed. When emotions such as love and hope instantiate the third type of life condition, Wittgenstein sometimes speaks of them as "emotional attitudes" (*Gemütseinstellungen*, e.g., *RPP* II, 152). This shows, first, that the second category is not identical with emotions and moods, though most conditions of this category will be instances of them; and, second, that some emotions and moods have cognitive features.

The fourth category is actions. Until now, I have been writing as if all ways in which things stand and are going for people are psychological conditions. But the preceding analysis of life conditions applies equally well to actions. Like psychological conditions, actions consist in (or, in the case of omissions, in the absence of) particular behaviors in particular circumstances. This indicates, notice, that behaviors must be distinguished from actions. A behavior is a bodily doing or saying, a type of action. Actions are either bodily doings or sayings or something a person carries out by way of performing a bodily doing or saying in a specific circumstance. For instance, Jane's asking a question might consist in her saying some words after raising her hand in class, whereas praising her child might consist in stroking, smiling, and saying encouraging words after it has accomplished something. Her speaking, raising, stroking, and smiling are actions of the behavioral sort, and they constitute the further actions of asking and praising in the circumstances in which they are performed. This sort of analysis is familiar in contemporary theory of action, where it standardly takes the following form: The performance of any action consists in the performance of a "basic action," usually involving the body, which in the circumstances amounts in one way or another to carrying out that action.[18]

The key point in the present context is that the "consists in" relation that holds between phenomena of life and psychological affairs also holds between these phenomena and actions. Indeed, the performance of an action consists not only in a bodily doing or saying (basic action), but also in sensations and images that accompany that behavior. Jane's asking a question *on that particular occasion* consists also in the queasy feeling in her stomach that accompanies her speaking, just as her running the 100-yard dash later in the day consists also in the image of victory faintly hovering before her mind as she dashes down the track. Behavioral expressions are of course more characteristic and central to action: Generally speaking, the concept of an action answers to the doings that appear in the phenomena of life (see *PI*, 620, substituting "the happening in the appearance" for "the phenomenal happenings" as a translation of *das Geschehen in der Erscheinung*). But inner expression is also relevant to a full appreciation of Jane's asking the question or dashing down the lane. In addition, some actions consist primarily in images, namely, imaginings (see *PI*, 615). Here, conversely, the state of mesmerized attention in which the body is held while imagining something is also constitutive of the activity. It is worth noting that Wittgenstein's remarks on willing and on voluntary versus involuntary behavior similarly make reference to appearances of life in particular circumstances. I should further point out that although behaviors (bodily basic actions), like mental conditions, are continuously expressed while performed, this need not be true of actions. Even campaigning for the presidency is not continuously expressed in a candidate's behavior throughout the campaign. And, finally, also as with mental conditions, particular behaviors and patterns thereof are associated with many actions, for example, reading (*PI*, 156) and driving a car.

Other reasons for treating actions as conditions of life press. For instance, the contexts in which appearances constitute particular actions are the same as those in which they constitute particular states of consciousness, emotions, and attitudes. What's more, the web of existing life conditions with which any new expression must mesh includes actions (a fact Davidson emphasizes).[19] Finally, when we ask someone what he is doing, just as when we ask him what he believes, intends, feels, hopes for, or loves, we are asking him to tell us something about his life, about what is going on or how things stand with him. All in all, it is doubtful that there is any significant division between the realms of mind and action. This is why I have heretofore been writing "mind/action" instead of "mind and action."

To sum up, avowals and attributions of life conditions of a particular category inform us about the dimension of life associated with that category (see Table 1). For being in a condition of a particular

Table 1. *Conditions and dimensions of life*

Condition	Dimension
1. States of consciousness	What one is explicitly aware of
2. Emotions and moods	How it is going with one
3. Cognitive/intellectual conditions	One's stances and attitudes
4. Actions	What one is doing

category articulates how things stand in one's life with regard to the corresponding dimension or aspect.

We can now consider Wittgenstein's notion of the inside of a person's life. Traditionally, the "inside" of a person was countenanced as that person's mind, the interior realm of representations, feelings, desires, and other mental denizens to which that person alone has direct cognitive access. To speak of an inside was to emphasize the privacy of this realm, in contrast to the publicness of a person's "outside," her behavior. In Wittgenstein, life has two types of inside. The first is the interiority of sensations, feelings, and images. Like the Cartesian inside, this inner is private. Unlike the Cartesian inside, it is not composed of objects identified on the basis of their features. The privacy of sensations and images, a person's having privileged access to them, is nothing more than their being appearances *to himself* of his own life conditions.[20] If I am in pain, I experience this in a way no one else can, namely, I feel it; another can but observe my behavior and empathize with and tend to my situation. But this sort of privacy does not mean that I alone can attest to my condition. Others can with equal authority pronounce on this on the basis of my pain behavior, for pain behavior is an appearance of being in pain as much as pain is.[21]

The second sort of inside is conditions of life generally (see *RPP* II, 604, 663). To cite a previously mentioned quotation: "And if the play of expression develops, then indeed I can say that a soul, something *inner* is developing" (*LW* I, 947). Life conditions compose the inside of life in the sense of the content of life, what is going on and how things stand in life. It is this inside that someone else can better understand than the person of whose life it is the content. A person does experience her conditions in a manner no one else can and can often report her conditions when paucity of behavioral expression precludes others doing this. But friends, associates, and therapists, on the basis of speech acts, the continuous manifold of sometimes "imponderable" behavioral expressions, and their occasions, are sometimes better able than the person herself to specify how things stand or are going for her. There is nothing clairvoyant or mysterious about this state of affairs. It simply reflects differences among people in experience, attention, sensitivity, and judgment. In this regard, to know what is going on in someone is simply to understand her (cf. *RPP* II, 714).

In conclusion, for Wittgenstein mind/action is how things stand and are going for someone in ongoing life. A human life, moreover, unlike a Cartesian thinking substance, is a publicly transpiring process. So mind/action must somehow be present in the public realm of sense experience. It is so via the play of expressive bodily doings and sayings. This holds equally true of actions, which we tend to think of primarily as entities in public space, as of having sensations, which we are too prone to view as a private affair. Moreover, since how things stand and are going is present in the world via the play of bodily phenomena, a person can see others' conditions by observing these phenomena. Perceptual access to mental and cognitive conditions will seem incredible only to those who insist that these conditions are tucked away in a private substance or realm called the mind (or to those who insist that the real objects of perception can only be physical objects or sense data).

Mind can no longer be thought of as a site or place housing the various attributes and functions called "mental." "Mind" now comprises but a collection of conditions marked by ascriptional asymmetries. But while no mind unifies the variety of attributes and functions, there is a body that expresses life conditions in common. Bodily doings and sayings, and bodily sensations and feelings, are the medium in which life and mind/action are present in the world. The body thus offers itself as an ersatz seat of mental unity. This newfound prominence of the body demands extended discussion.

The Expressive Body

By way of the body, mind is present in experience. Bodily doings and sayings are the realities in which psychological conditions consist in particular circumstances. This does not, however, reduce life conditions to bodily activities. Bodily activities are simply what there is in the (experiential) world to such conditions. The conditions themselves maintain, so to speak, a notional existence beyond the phenomenal expressing them. Life conditions are aspects of how things stand with a person's existence, and things standing someway for someone is a state of affairs not reducible to what manifests it in the world. But what a person is about, this state of affairs, is also not a thing or phenomenon. It has no substantial being over beyond its expressions. So it also is not causally responsible for its expressions, and cannot be identified with anything, for instance, bodily states, which are related causally to bodily activities.

In understanding this conception, it is helpful to mention an expression of Medard Boss, the German phenomenological physician. As a student of Heidegger, Boss dispenses with the picture of mind as a realm or thing. What had been gestured at with the notion of mind is instead conceptualized as human existence. Human existence, moreover, exhibits a number of structures, including "mineness" (the fact that existence is always my, i.e., someone's, existence), the threefold structure of temporality (thrownness, falling, and projection), spatiality, and being-with (coexistence). Boss goes beyond Heidegger in adding to this list "bodyhood," the human body's "bodying forth" a person's current way of being:

Human bodyhood is *always* the bodying forth of the ways of being in which we are dwelling and which constitute our existence at any given moment. . . . [a person's] bodyhood occurs exclusively as the bodying forth of his existential dwelling amid the beings that address themselves at any given time to his perception and require of him an appropriate response.[22]

For Boss, every manifestation of existential ways of being is bodily. I would claim that this is a Wittgensteinian position too were it not for unclarity concerning the bodily nature or status of mental images and words.

By "ways of being," Boss means modes of existence that are expressed in everyday language by the words I have interpreted as designating conditions of life ("being in pain," "fearing," "believing," "running," and the like). This (Heideggarian) notion of a way of being and the Wittgensteinian notion of a condition of life are more or less the same.[23] For Boss, a person's ways of being are bodied forth in behavior and in states of the physical body such as stomach pains. This idea is remarkably similar to Wittgenstein's, since behavior for Wittgenstein is the primary locus of expression, and most inner phenomena in which conditions of life consist are bodily sensations and feelings. Boss's expression "body forth" is particularly propitious, both because it captures the body's making present and giving others to understand how things stand in a person's life and because it declines to identify or reduce ways of being to the bodily activities and states manifesting them. It also retains a feel for life as a stream. As we live and enter and are claimed by various conditions of life, we always press forward in and as our bodies, behaving and speaking in ways that express these conditions.

Until now I have described the arena of expression as doings, sayings, and sensations. Making this account more precise requires differentiating three regards in which the body expresses conditions of life. These three regards correspond to three dimensions of bodyness, of what it is for a person to be a body. Recall that a person, most centrally, is that to which conditions of life as well as physical attributes can be ascribed. That life conditions are so ascribable means that both behavioral and inner bodily episodes are too. To be a body, consequently, is to be able both to perform bodily doings and sayings and to experience bodily sensations and feelings.

Of note is that being a body is but one component of embodiment, one that emphasizes the lack of experiential and conceptual disunity a person has with her body in normal circumstances of acting and experiencing. A second component of embodiment is having a body.[24] Having a body is made evident in situations of breakdown, malfunction, discomfort, and incompetence, where the fact that one is a body manifests itself explicitly. In cases of injury, disease, failure to achieve goals through physical effort, the gazes of others, and the acquisition of bodily skills, a person is forced to confront and deal with her body. Such experiences may also underlie her recognition of a distinction between herself and her body. Although this body is hers, she is not identical with it. Rather, she has it.

The first modality of bodily expression is manifestation. The body is a manifesting body. Recall that mental conditions are continuously expressed. Their existence implicates the body as the medium of manifestation for the instantaneous, moment-to-moment, ebbs and flows of human life. The conditions currently gripping and configuring people's existences are ceaselessly made present in the world through bodily activities and feelings.

The body is, second, a signifying body. Not all a person's conditions are continuously expressed in her behavior. A person can hold attitudes and positions, for instance, even when unexpressed. Sometimes, accordingly, Wittgenstein labels believing and knowing (though not all cognitive conditions) "dispositions." This term is risky, since many people think of a disposition as the state of a momentarily inactive apparatus that determines how the apparatus will respond to certain future stimuli; and, as we have seen, Wittgenstein opposes an apparatus picture of mind (in this regard, see *PI*, 149). A disposition in his sense is more like certain behavior being expected from someone in certain circumstances (cf. *LW* II, p. 9d). (This is, of course, a weak sense of "disposition.") The body, then, is a signifying body in the sense that its activities can signify (predominantly to others) that a person can be expected to perform certain other bodily doings and sayings in specific circumstances.

The point of saying that the body "signifies" is that cognitive conditions usually differ from mental ones in lacking characteristic expressions and not being visible in their expressions. While the body palpably manifests emotions and consciousness, its expressions of cognition and intellection are less specific and forceful. Which cognitive condition(s) a given phenomenon expresses is more indefinite and less socially specified. In the face of a play of behavior and occasions, consequently, an observer will often have to ponder what occurred and infer which cognitive conditions were expressed. This requirement only sometimes applies to mental conditions, which can often be directly seen in or read off behavior. What's more, to the extent that cognitive conditions not only lack characteristic expressions but are not expressed at all, ascribing them must attend to such considerations as what is generally believed in the person's community or group, what beliefs would be expected of a "rational person" as understood in that community, and what information on a particular topic is generally available or specifically accessible to the individual concerned. This means that the attribution of cognitive conditions sometimes relies on reconstructed webs of belief and knowledge, which are not directly and immediately expressed by the sayings and doings of the person to whom they are attributed, but are instead ascribed to him because of his social positions, his community, his biography, and so on. In other cases, of course, cognitive attributions articulate what a person's behavior "directly" signifies about him. These cases, however, often presuppose detailed knowledge of the person's life. Hence, the attribution of cognitive conditions requires, at one and the same time, both far more specific knowledge of people and their situations, and far more familiarity with sociohistorical contexts, than the identification of mental conditions does. Claims about cognitive conditions, as a result, are usually more uncertain than assertions about mental ones. The immediacy of the manifestation of mentality is replaced by the inferentiality of the signification of cognition.

The body is, third, an instrumental body. It is through the performance of bodily actions that the performance of other actions is constituted or effected. One way of seeing this is to relate the foretold story about basic bodily actions to contemporary views of Aristotle's practical reasoning or, better, Heidegger's notion of signifying (*bedeuten*).[25] In Heidegger, signifying is that which determines which action, of all those a person is able to perform, is the one she does perform in a given situation at a given moment. It consists in "signifying chains" (my expression), which stretch from conditions of existence for the sake of which the person acts (ends) to specific actions, which are specified as what to do at this moment in this particular situation for the sake of that condition. Such chains can be short or long. Moreover, they often, though not always, pass through series of stages, in each of which an increasingly more specific project, task, or action is singled out on the basis of the actor's beliefs, moods, situation, and desires (etc.). So, for instance, acting for the sake of winning Michael might signify to Teresa, given her beliefs and moods, buying him flowers, which project in turn might signify to her, given her beliefs and situation, driving to the nearest florist, a task which in turn might signify, given the mechanics of automobiles, finding the car keys – so she stands up and begins to look about the room. Intermediate stages of these types need not, however, be present. A person might, for instance, order chocolate ice cream simply (directly) for the sake of enjoying it.

Signifying always terminates in a particular action, which is what now is to be performed for the sake of an end on "active service." Although Heidegger never discusses this, the performance of practically any action consists in the execution of a bodily one. So once signifying has singled out an action, the person therewith carries out a bodily action that, in the circumstances, effects the performance of the signified one. The transition from signification to performance is in the normal case automatic. Each person has an indefinite repertoire of bodily actions that he or she can carry out automatically, that is, spontaneously and without thought. Whereas the ability to perform some arises from biological wiring, the capacity to perform most is acquired through social training and upbringing. Signifying terminates at the specification of an action that can be carried out by the performance of one of the actions contained in the bodily repertoire. So the termination of signifying is at once the performance of a bodily action that constitutes the signified action in the circumstances involved. Bodily performance is unceasingly instrumental in the achievement of a person's ends.

In saying this, I do not mean that the body is an instrument that some disembodied will or intelligibility takes hold of, as it were, and exploits to achieve its ends. Signifying must not be understood on the order of an immaterial mind that controls the body as a captain steers a ship. A person *is* his or her body. Once it is signified to someone that such and such an action is to be carried out, he therewith performs the bodily action that constitutes the signified one. For it is the same person signified to who, *in being his body*, carries out bodily activity. There is no need for him to seize, occupy, or activate his body in order for bodily activity to occur. (Z, 586, suggests that Wittgenstein has a sense for this.) An instrumental body does not imply that the body is an instrument.

Now, since a given bodily performance can instantiate multiple dimensions of expression, these dimensions are not mutually exclusive. Chanting at a political rally, for instance, can simultaneously signify political beliefs, manifest joy in collective political activities, and constitute making a statement against government policy. Most bodily doings in fact manifest mental conditions and signify cognitive ones in addition to constituting actions. On the other hand, not all bodily activities that express mental and cognitive conditions constitute actions. There is a range of behavioral goings on, for instance, nervous fidgeting, that are not bodily actions and whose occurrence does not constitute the performance of further actions, but which nonetheless manifest mental or cognitive conditions (e.g., anxiety or disbelief).

Incidentally, the distinction between being and having a body cuts across the three dimensions of bodily expression. With a range of bodily expressions at her disposal, a person is usually at home in her body, automatically expressing mental and intellectual conditions and carrying out actions through effortless and unconsidered bodily performance. She is unproblematically a body. Breakdowns in any dimension of bodily expression, however, can impress upon her a distinction between her and her body. Examples of such disturbances are debilitating depression, uncontrollable crying or laughter, chronic or acute bodily pain, paralysis, nervous ticks, and the effort required in learning complex, subtle, and coordinated activities such as piano playing or skiing. When such disturbances occur, the body either ceases or fails to express certain conditions of life spontaneously. Expression modulates, adopting distorted bodily forms, migrating largely into inner phenomena, or forcefully acquiring new configurations. With the frustration, encumbrance, and deformation of expression, the unquestioned coincidence of a person and her body is broken.

This Wittgensteinian vision of an expressive body strongly resembles Judith Butler's performance theory of gender identity. Like Wittgenstein, Butler rejects the notion that a person (gender, or sex) is a substance, or metaphysical substrate, whose identity lies in the continuing possession of specific "inner" properties. Again like the position outlined here, she conceives of personhood and identity as statuses or positions instituted in social practices: Only within the context of Foucaultian regulatory practices can there occur persons with specific identities. Butler further analyzes personhood and gender as (bodily) do-ings instead of be-ings. More precisely, they are standings attained through behavior (as opposed to attributes of a substance or realm). Gender identity, in particular, is a corporeal style, instituted in stylized repetitions of acts.[26] Butler calls this a "performance" theory because the performative reiteration of a specific (normalized) style of behavior signifies that a person is of such and such a gender. (It also signifies the illusion that there is a "gender core" responsible for the acts. Indeed, the ideas of a soul and of an inner realm are further fabrications likewise signified by acts and gestures.[27])

On the account outlined here, bodily activities manifest, signify, and constitute conditions of life. In expressing mind/action, these activities at the same time institute individuals. For in expressing particular conditions, bodily activity *eo ipso* establishes *that there is* someone in them. The existence of an individual amounts, above all, to there being a body that, in expressing specific conditions, thereby establishes that there is someone who performs these actions and is in these mental and cognitive conditions. This situation is borne out by the fact that even though bodies express conditions of life, it is to persons – not bodies – that conditions are attributed. Furthermore, insofar as a person's identity centers upon her conditions, bodily activity also establishes that there are individuals with particular identities. All these theses closely parallel Butler's idea that bodily activity "constitutes" gender identity. As we shall see in the following chapter, individuals ultimately exist by virtue of the incorporation of human bodies into social practices wherein they become expressive bodies. Similarly, gendered subjects come, for Butler, to exist through the incorporation of bodies into discursive practices whereby bodily activities (and features) come to "reiteratively cite" and thus materialize the gender norms carried in those practices.

I have been describing the "outer" domain of expression as bodily doings and sayings. Doings and sayings are two forms of behavior, a unity-in-difference reflected in Wittgenstein's use of the words *Ausdruck* and *Äusserung*. Whereas *Ausdruck* is consistently translated as "expression," *Äusserung* is usually rendered either by "expression" or by "utterance." This is notable, since the standard (though not exclusive) translation of *Äusserung* outside philosophical contexts is "utterance." "Expression," or even "manifestation," is forced upon the translator, however, whenever what is referred to as an *Äusserung* is not a speech act but some other form of behavior. Similarly, Wittgenstein uses *Ausdruck* to refer to both behavioral performances and linguistic words and phrases. Neither word, consequently, lines up straightforwardly with either type of expressive behavior.

By "bodily doings" I mean all behavior that is not a speech act. For my purposes, I will define a speech act as any behavioral episode in which something is said.[28] So defined, speech acts, or bodily sayings, need not involve language, at least in the sense of verbal or written words. Shakes of the head, waves of the arm, winks, and so on say something and thus count as speech acts as much as utterances of words do.

That bodily *doings* – from walking and caressing to stomping and shaking, from crying and laughing to frowning and sighing – express conditions of life is an obvious fact requiring no discussion. Of greater complexity are the expressive powers of bodily *sayings*, especially linguistic ones. For one thing, a person can be in certain conditions of life only if he is capable of using words. Wittgenstein implies, for instance, that only a creature who can talk can hope (*PI*, p. 174; cf. *LW* II, p. 67c). Linguistic acts also make plain the multidimensionality of behavioral expression. Avowals of life conditions, for instance, sometimes express other conditions more emphatically. If a person cannot get the hang of a parallel turn and tearfully moans "I can't do it," this avowal of inability is more poignantly an expression of frustration. Uttering certain words about oneself in a particular tone on a given occasion can express something about oneself more significant than what is said.

Self-ascriptions can also, of course, express the ascribed condition. An utterance of "I hope he'll come" might just as much express hope as does an utterance of "He'll come, won't he?" (or repeatedly rereading his letter to glean whatever information possible). Wittgenstein repeatedly returns to the fact that nonavowals can be substituted for avowals to suggest that sometimes such avowals as "I hope he'll come" do not so much say something about oneself as manifest one's condition (hopefulness).[29] (The difference lies in whether one is telling someone about oneself or simply being oneself.) That is to say: Despite their surface form, first-person uses of words that designate life conditions are at times expressions of and not reports of such conditions, which act as signals for others. But this is not always so. Uttered in response to a query about my hopes, "I hope he'll come" is a report about my condition and not an expression of it.

But not only are "avowals" sometimes expressions and at other times reports, they can also be descriptions or explanations. And that one and the same first-person use of a mental locution can be a move in any of the sometimes mutually exclusive language-games of expression, description, report, and explanation, holds for locutions that designate any of the four types of condition distinguished earlier. Third-person uses, too, enter the same variety of language-games. But the greater complexity of first-person usage is evidenced by the fact that first-person forms of such verbs as "to believe," "to know," and "to hope" at times do not function primarily to describe, explain, report, or express life conditions, but to realize such interpersonal achievements as assuring others that something is the case, indicating that something can be relied on, averring that something was learned about long ago, suggesting that something is not for certain, or putting pressure on someone to do something (e.g., *OC*, 176; *RPP* II, 2; *Z*, 406). This fact repeatedly led Wittgenstein into suggestive but inconclusive investigations of the differences between first- and third-person uses of such verbs as "to believe" and "to know."

In addition to manifesting mental states, speech acts can also signify cognitive conditions and constitute further actions. All sorts of statement, and not just first-person utterances, signify a person's desires, doubts, and intentions and let others know what she does and does not understand. As John Austin and John Searle made clear, furthermore, speech acts are also often performative, meaning that an actor, in carrying out a speech act, performs a further act. A classic example is a parson making a couple husband and wife by uttering "I pronounce you man and wife." In any event, speech acts, like all behavioral phenomena, can, often at one and the same time, manifest, signify, and effect different life conditions.

In treating speech acts as behavioral performances, I mean only to stress that they, like nonlinguistic behaviors, are directly carried out bodily. I do not follow those theorists (such as C. K. Ogden and I. A. Richards in one tradition and Maurice Merleau-Ponty in another) who aimed to reduce, for instance, linguistic meaning to behavior. Nor would I interpret Wittgenstein's dictum, that the meaning of a word is in a large number of cases its use in the language, as implying that meaning is reducible to behavior. Still, Wittgenstein's remarks on speech acts emphasize bodily skills at the expense of cognitive abilities. A moment's reflection on this will deepen appreciation of language use as a form of bodily performance.

Wittgenstein analyzes understanding language as the ability to go on using words in ways intelligible to others. As disclosed by his observations about games and family resemblances, however, this ability defies explicit presentation. The flip side of the famous claim in the *Philosophical Investigations,* that the various phenomena called games are linked only by family resemblances, is that what counts as a game has no "frontier." For every putative frontier (i.e., every attempted definition of "game") proceedings can be thought of that do not fit the definition but would still be called games.[30] This shows, though Wittgenstein does not draw this conclusion explicitly, that no formulation, fashioned either by a speaker or an observer on the basis of a finite number of uses of the word "game," will be able to cover all possible acceptable uses of it. Since the open field of acceptable uses arises from the understanding of the concept of a game, it further follows that formulations cannot exhaustively capture this understanding. Ventured definitions do capture something of it. That, however, what is uncaptured is inexhaustible is shown by the perpetual imaginableness of proceedings that contravene proffered formulations but still count as games on the basis of the understanding. Hence, the practical ability to use the word, to apply it in novel and unanticipated circumstances, outruns any formulation that purports to capture the ability and to delimit the usage. As a matter of logic, moreover, indefinitely many rules can adequately represent any finite number of uses. So the past use of the word is no more in accordance with any one than another of these rules.

For these two reasons, it follows that the use and understanding of the word "game" cannot be fully presented in the form of explicit propositions. This predicament is reflected in the facts that children usually learn to use words without being given explicit formulations of their meanings (e.g., *OC* 95, 140; *LW* I, 968) and that even when rules are brought to bear, examples too are needed. (The practice, Wittgenstein writes, must "speak for itself" [*OC*, 139].)[31] This result, moreover, is perfectly general. Rules can capture the use of only those words introduced through explicit rules (and arguably not even these), and more or less all of natural language is not such.

It is important to add that these considerations, it seems to me, apply to human activity at large: Whatever abilities to go on intelligibly out of which people nonverbally proceed likewise cannot be exhaustively formulated in words. This is suggested by the fact that the same infirmity of formulation that Wittgenstein spots with regard to linguistic understanding also infects the know-hows pertaining to nonverbal activity: The abilities to go on nonverbally that are evidenced in both actions and judgments of intelligibility and correctness cannot be adequately reproduced in linguistic formulations. At best, the intelligibility of proceeding in particular ways in specific situations can be spelled out in detail, and general rules of thumb about how to proceed in certain sorts of situation devised. These rules of thumb, however, even more than putative definitions of "game," fail to anticipate how people go on sensibly in all situations in which they (prima facie) apply. They do not, therefore, delimit the capacities to go on that are evidenced in people's intelligent coping in these situations.[32] Knowing how to act cannot be laid out in explicit formulations.

Whether one is well advised to continue averring that the use of a word (and behavior more generally) is "rule governed" remains a separate question. Perhaps word usage is governed by rules that cannot be fully formulated. Although different passages in his late writings suggest that Wittgenstein does or does not view language use as rule governed, one paragraph in particular (*PI*, 82) highlights the futility of retaining talk of rules. Here Wittgenstein writes that when neither language users nor those observing them are able to formulate rules that adequately capture the use of a word, "What meaning is the expression 'the rule by which he proceeds' supposed to have left to it here?" The moral I draw is that, in the face of the inability to formulate adequate rules, retention of the belief that language use is cognitively guided, in the specific sense of conforming to or being determined by rules (meanings, signifieds, concepts, or forms), is more likely to mislead than enlighten. It is far better to renounce the idea that rules and meanings guide or determine language use and instead allow future reflection to be guided by the idea of unformulable understandings – wherever this may lead. From this perspective, rules and meanings are really only ex post facto attempts to recoup past usage. In slogan form: It is what we do, how we go on, that determines the rule, not vice versa. I might add that the familiar claim, that it is appropriate to speak of language use (and behavior generally) as rule governed when it makes sense to distinguish between semantically or normatively right and wrong ways of speaking, does not shore up the idea that language use is governed by rules that cannot be fully formulated.[33] For in conceding the unformulability of rules, this claim either maintains that *all there is* to rule following is the existence of right and wrong ways of speaking and acting, in which case talk of rules adds nothing; or it aims to *explain* the existence of these right and wrong ways by reference to unformulable rules, in which case rule talk is unnecessary since people's understanding of language suffices to explain this. Talk of unformulable rules is at best a potentially misleading way of referring to this understanding. Similar remarks apply to the idea that language use is governed by rules because to follow a rule for the use of an expression is simply to use the expression as it is ordinarily used.[34]

Mastering a language is not a matter of following rules and meanings, but of being able to go on using words intelligibly to others. The causal genesis of this capacity might be discovered, but its content is unpresentable. Indeed, on Wittgenstein's view, all human action and thought is underwritten by a repertoire of noncognitive (in the sense defined here) abilities to carry out bodily performances. And in this regard, his ideas converge with those of Maurice Merleau-Ponty on the sedimentation of bodily abilities:

I do not need to visualize the word in order to know and pronounce it. It is enough that I possess its articulatory and acoustic style as one of the modulations, one of the possible uses of my body. I reach back for the word as my hand reaches towards the part of my body that is being pricked; the word has a certain location in my linguistic world, and is part of my equipment. I have only one means of representing it, which is uttering it.[35]

In the work of Wittgenstein and Merleau-Ponty (as well as in that of the philosopher Hubert Dreyfus and the sociologist Pierre Bourdieu), the realm of ideal, guiding rules or meanings is replaced by a background of past bodily sayings. A person picks up the ability to perform bodily sayings intelligible to others by living through speech acts, by encountering, attending to, learning, and venturing them. Past bodily sayings, in forming the context of the acquisition of this ability, orient, but do not determine, how language will be acceptably used in the future. Instead of dictating the correct use of language, they provide individuals with the capacity to cope with novel situations linguistically, to employ language in the face of endlessly varying events.

Incidentally, the fact that language use lacks "cognitive" tracks provides one interpretation of Wittgenstein's assertion that "The new (spontaneous, 'specific') is a language-game" (*RPP* I, 164; *PI*, p. 224; cf. *RFM* IV, 23). Language use is a reaction to the world not pinned down by rules, meanings, past usage, ideas, or anything else. Outrunning determination, speaking and writing are in a fundamental sense spontaneous (which is not the same, of course, as being random or willful). Every speech act, moreover, is new since it occurs in and is specifically appropriate to a unique situation. Speech acts, consequently, are spontaneous behaviors that are appropriate specifically to their situation (and which often originate a sequence of speech acts and behaviors that is also specifically appropriate).[36]

To close this section and as a preview of the following chapter, I wish briefly to consider the social context of bodily expression. A human body becomes a manifester, signifier, and constituter of life conditions largely through social molding. All functional human beings, Wittgenstein believes, share certain "natural" bodily reactions that express conditions of life. The most prominent examples he mentions are reactions to bodily injury, such as crying when struck and tending to a hurt bodily part. What makes these reactions "natural" is that they are prelinguistic (*Z*, 541). This limited, biological level of natural expressivity forms the basis of an extended and extendable realm of culturally instituted expression. Through social training and learning, a person comes to augment natural expressions with more elaborate bodily and, in many cases, linguistic expressions that express the same conditions. Social training, especially in language, also enables a person to express a wealth of conditions that the natural bodily repertoire does not support at all (e.g., *PI*, p. 174). Eventually, a person comes to possess an extensive repertoire of possible bodily doings and sayings, which far surpasses his natural biological endowment and with which he can express a range of conditions of life exceedingly more extensive than the limited range of conditions accessible to him via that endowment. So the possession of an expressive body, and therewith of mind, is the product of social training and learning.

In the previous section, I mentioned that the range of life conditions open to people is tied to social practices. A correlative point is that the range of possible bodily expressions of these conditions is similarly a social product. A person, as just discussed, acquires an extensive repertoire of bodily doings and sayings through social training and learning. As we saw earlier, furthermore, just which conditions are expressed by particular behaviors in the repertoire is also largely a matter of cultural determination. (On the other hand, that mind/action is expressed by the body is an ultimate, nonsocially determined fact about human existence.) We can now add that the interests that determine which patterns of expression people notice in bodily activities and fix in their concepts (see earlier discussion) are social interests in one another's lives. Mind/action is a social institution, the body that expresses it a social product, and the interests that guide these processes interests in human coexistence.

Although the expressive significance of bodily activity is socially instituted, each person, in his bodily activity, exhibits a unique style: a unique subset of particular doings and sayings and unique ways of performing common ones. Bodies are not passive objects seized by social practices and molded into clones (or robots)[37] that perform stereotypical activities in common. Even natural reactions exhibit nuanced variation. A person's unique style of bodily expression evidences an active being who learns social forms but then expresses itself. The point of saying this is that, just as before we saw it is wrong to view the body as the instrument of an immaterial free will or intelligibility, so too is it wrong to conceive of it as the passive plaything of social forces. The expressive body cannot be locked into the traditional dichotomy of free will versus determinism.

Coda: What Is Mind?

I earlier credited Wittgenstein with helping to deunify and desubstantialize mind. The variety of mental attributes and functions traditionally united as properties of a substance or occupants of a realm are unmoored from their housing through its dissolution and transformed into socially instituted aspects of life marked by the peculiarity of ascriptional asymmetry. These aspects, however, find new unity through bodily expression. The body is the seat, or site, of mind *qua* common medium of expression for the disparate conditions of life. As we have seen, this medium is a socially molded multidimensional site of manifestation, signification, and effectuation where life conditions are bodied forth in the phenomenal world.

Mind, consequently, if we want to continue employing this substantive, and if we want further to contravene Wittgenstein's admonitions against formulations seeking generality, is the expressed of the body.

In this context, a phrase of Ludwig Klages comes to mind, that the body is the appearance of the mind and the mind the meaning of the living body.[38] The pertinence of this phrase is confirmed by the places where Wittgenstein muses that "The human body is the best picture of the human soul [*Seele*]" (*PI*, p. 178; cf. *RPP* I, 275–276, 280–282). In Wittgenstein's idiom, Klages' thought would be rewritten as: Bodily activity is the appearance of mind, and mind is the expressed of bodily activity. As we have seen, bodily behaviors, sensations, and images are the phenomena or appearances of mind/action, what there is in phenomenal reality to actions and mental and cognitive conditions. Correlatively, bodily phenomena, unlike, say, rocks, make present in the world something other than themselves, namely, mental as well as cognitive conditions. The living human body is inherently expressive: It manifests, signifies, and constitutes how things stand and are going in a person's life. Mind, consequently, comprises the ways of being, or conditions of existence, made present by the expressive doings and sayings of human bodies.

3

The Social Constitution of Mind/Action and Body

Philosophers and social theorists once found it unproblematic to conceptualize mind as a distinct and autonomous ontological substratum separate not only from the world, but also from the actions intimately connected with it. The first component of this conception to fall was the autonomy of the fundamental structures and contents of mind from social context. Hegel initiated its decline with the claim that individuals are constituted within Ethical Substance, in the language, institutions, and practices of a society. Although this conception subsequently became paradigmatic for an array of thinkers, these thinkers, including Karl Marx despite his materialism, continued to accord mind distinct existence as a realm or thing.

Cartesianism, understood as the doctrine that mind possesses such distinctness, remained a vital and resilient conception in the late nineteenth century and first few decades of the twentieth century. In philosophy, for example, some version or other of it remained firm for a slew of thinkers, from the analytic innovators Gottlob Frege, Bertrand Russell, and G. E. Moore to such philosophers of consciousness as Ernst Mach, Henri Bergson, Edmund Husserl, and the early Jean-Paul Sartre, from the Vienna School theorists Morris Schlick, Otto Neurath, and Rudolf Carnap to the pragmatist William James and neo-Kantians such as Heinrich Rickert. These theorists were seemingly oblivious, moreover, to the dependency of mental contents on social context. Sigmund Freud and Georgi Lukács also continued to treat mind as a distinct realm, even though the one theorized a socially molded psyche and the other extended Marx's vision of the class determination of consciousness and knowledge.

In the 1870s, Friedrich Nietzsche pointed toward the possibility of critiquing the ontological distinctness of mind in conceiving the inner as a tissue opened by the inward discharge of the instincts to freedom and power (which turn inward because society frustrates their outward expression). It was not until the 1920s and 1930s, however, that the critique emerged. Heidegger's *Being and Time,* the pragmatist theories of Charles Cooley, George Herbert Mead, and John Dewey, and the behaviorism of John Watson spearheaded this movement. The idea that mind is socially constituted also received fresh impetus during this period. Oswald Spengler, for instance, notoriously claimed that the nature of mind is relative to cultural type and historical moment. Elements of social constitution also appeared in the work of the Frankfort School theorists.

Wittgenstein's late work was thus written at a time when the distinctness of mind as realm or thing had come under direct assault and belief in its autonomy was steadily waning. His remarks, however, do not easily fit into the intellectual history of the gradual acceptance of mind's social nature. On the one hand, Wittgenstein read few books, especially those of contemporaries, and the development of his version of the social perspective is best referred to the dynamics of his own intellectual evolution.[1] On the other hand, subsequent articulations of this *Weltanschauung* largely ignored his writings. The extensive discussions in analytic philosophy during the late 1950s and 1960s that centered on his work, especially the so-called private language argument, indeed concerned the social constitution of mind (though analytic philosophers more often conceptualized these discussions as arguments about the public nature of language and consciousness alone). During those years, however, Wittgenstein was generally stylized as a philosopher of language, knowledge, mathematics, and sensations, of interest primarily to Anglo-American philosophers and of little use to social theorists in general or those concerned with social constitution in particular. This insulation began to erode in the 1970s as the German philosopher Karl-Otto Apel, the French theorist Jean-François Lyotard, and the English sociologist Anthony Giddens either put his concept of a language-game to constructive use or took inspiration from his analyses of what it is to follow a rule. In more recent times, David Bloor and Chantal Mouffe have furthered this process of assimilation.

Wittgenstein's remarks harbor an incisive and comprehensive analysis of the social constitution of mind/action. This account not only pertains to all dimensions of mind/action, but is part and parcel of his denial that mind is a distinct realm or thing. Wittgenstein's texts are thus among the first to outline a positive account of the social nature of mind/action that denies both its autonomy and separateness *qua* realm. As we shall see, this analysis also highlights the role played by the body in the constitution of mind/action in a way that anticipates contemporary interest in the social molding of the body.

Before turning to this account, I want first to make a remark (primarily for philosophers and Wittgenstein scholars in particular) about possibility and actuality. Wittgenstein often characterizes his intellectual journeys as conceptual investigations that aim to clarify concepts, as opposed to hypothesizing inquiries that seek to explain phenomena causally. Although this opposition does not imply that his conceptual elucidations either neglect actual phenomena or fail to illuminate them, an important component of his approach to the clarification of concepts is imagining nonexistent phenomena (i.e., human practices). A more explicit and thereby surer understanding of a concept is secured by judging whether it applies to or within imagined practices that deviate in certain ways from the familiar phenomena of our own lives to which the concept is actually applied. The possibility and character of concepts that depart either slightly or greatly from our own are likewise probed through the device of imaginary practices. Wittgenstein, as a result, is sometimes described as a philosopher of possibility, as opposed to one of either necessity or actuality.

Wittgenstein, however, is also intent on pointing out facts of "natural history," indeed more keen to do this than to deny that he is doing natural history (e.g., *PI*, p. 230). Moreover, the general facts about human lives, bodies, and social contexts that are revealed by his remarks, and about which he occasionally comments, in the first place concern how things happen to be with human beings. I do not refer here to the "general facts of nature" that Wittgenstein sometimes asks his reader to imagine altered (e.g., that the colors of things generally remain constant and are not continuously changing). I mean, instead, certain aspects of human practices, language use, and individuals that seem to me, inter alia, to embody an account of the social constitution of mind/action. That these aspects are in the first place features of human actuality does not gainsay their also characterizing the imaginary cases and practices on which the philosopher's appreciation of both particular concepts and concepts in general cuts its teeth. Perhaps some of these aspects would also have to hold of anything if it is to count as a human practice or life. But not all these aspects are thus (we can imagine, for instance, that children suddenly begin speaking one day without prior training or exposure to language); and it remains true that we know of them only from our own and other extant practices. Thus, Wittgenstein is also a philosopher of actuality, who has much to say about how human life actually works.

It is as a philosopher of actuality, of natural history, that I am concerned with him. Even if such and such a phenomenon is a necessary feature of any human practice whatsoever, I will be concerned solely with its natural historical role in actual human life. How, consequently, human life must or cannot not be reasoned or imagined to be is of no interest. Logical possibility and impossibility, too, lie outside the scope of the following study. And this means, as must be already evident, that the current study is not particularly concerned with elucidating concepts and charting their possibilities. It seeks instead to comprehend the social matrix as part of which concepts actually function.

The Formation of the Expressive Body

In this section I sketch Wittgenstein's picture of the progression of a human being from infancy to functional maturity, and from there to the novelties and extensions of intelligibility effected by a few. This sketch will be largely familiar to Wittgenstein scholars and to analytic philosophers conversant with the debates that were occasioned either by Wittgenstein's work or by interpretations thereof in the late 1950s and 1960s. I hope to paint this picture with new emphases, however, focusing in particular on the expressive body. I do not suppose, furthermore, that the general thesis outlined in this section, that expressive bodies are molded within social practices, is particularly controversial. The point of the current section is simply to fill in that component of the account of the social constitution of mind/action lurking in Wittgenstein's remarks that concerns the social molding of the body.

A key element in this account is the notion of a reaction (*Reaktion*). A reaction, for Wittgenstein, is a spontaneous behavior. Spontaneity, however, does not mean randomness, as it often does in physical contexts. It does not imply that the behavior concerned is either uncaused or without reason. It means simply that behavior is unreflective, or unconsidered. An action is unreflective when it is not preceded or accompanied by explicit reflection or consideration about what to do. The emphasis here lies on "explicit." Spontaneous behavior happens immediately (*unmittelbar*), without the intervention of conscious thought. This does not mean that the behavior involved can't be thoughtful, intentional, and even deliberate, only that its being such does not consist in a process of conscious thinking or deliberation that precedes or accompanies it.[2] The lion's share of any human being's behavior is unreflective in this sense. Moreover, central to what it means to say that a doing or saying is part of a person's bodily repertoire, in the sense introduced in the previous chapter, is that a person is able to perform it spontaneously. Such behavior is, in Wittgenstein's words, "next to hand" (*nächstliegenden, PI*, p. 201). When a reaction occurs in a context marked by strong regularity in behavior, Wittgenstein sometimes describes it as what someone does as "a matter of course" (*selbstverständlich*, e.g., *PI*, 238).[3]

This notion of reaction clearly differs from the notion of response that figures in behaviorism's concept of stimulus–response. In the first place, reactions are not always responses to a stimulus. They need not be responses to anything at all. The spontaneity at the core of reactivity bespeaks activity and self-energization, in the sense in which Aristotle said living beings possess their principles of motion within themselves. It connotes self-generating movement into the world instead of stimulated response to something. Even if a stimulus is conceived as a movement-inducing occasion instead of a mechanism-releasing spark, the therewith coordinated conception of response still fails to correspond to Wittgenstein's notion of reaction. Reactions are always occasioned, but this means simply that they occur in particular circumstances. Features of those circumstances do, moreover, sometimes induce reactions. But reactions, as self-activating interventions in the world, can occur even if nothing induces them. The world does not need to solicit them in order for them to take place.

Thus, second, there is nothing mechanistic about a reaction. Simplistic Pavlovian behaviorism pictured stimulus–response circuits as deterministic setups. If a given stimulus occurs, the circuit guarantees a particular response. More sophisticated behaviorisms such as that of B. F. Skinner built flexibility into stimulus–response circuits, such that a given stimulus can be responded to in a variety of ways. But causality still pervades this picture since the particular responses that do occur are produced by complicated webs of circuits established through public reinforcement schedules. Wittgenstein's notion of reaction, by contrast, does not imply anything in particular about production. Although, as I will discuss, reactions are probably caused by neurophysiological bodily pathways, this is an additional and hypothetical thesis irrelevant to the definition of a reaction. A reaction is a spontaneous, unreflective, behavior, whether and however it might be caused.

Reactions, moreover, can be either "behavior" or "action," in the contrasted senses usually given these terms (roughly, unintentional versus intentional movements-doings). Some reactions are instinctual and others are wrung reflexively out of us by, for example, the incisions of doctors and pests. These behaviors are spontaneous, unreflective reactions as much as the following actions can be: feigning as one begins to drive toward the basket, glancing admiringly at a landscape, giving way to oncoming traffic, extending greetings to a neighbor in the street, and forcefully expressing disagreement with a speaker's point. The "instinctual" behavior of the three-month-old, the unintentional behavior of the adult, and many of that adult's intentional actions as well are equally reactions. (As we shall see, adults also at bottom continue like infants to react in ways for which there are no intentions or reasons.) Reactivity is a baseline feature of the human condition that adopts a variety of forms.

Emphasizing the ubiquity of reactions brings to the fore that reactivity underlies all proceeding on the basis of explicit considerations, formulations, rules, and symbols. Spontaneous, situationally attuned ways of reacting always run ahead of the ability of conscious thought to master and direct them; and insofar as conscious thought does govern human activity, its ability to do so rests on the continuity of underlying reactions. This description of action bears likeness to Nietzsche's idea that consciousness floats on a neurophysiological hormonal nonconscious that governs human existence and is impervious to direct conscious and reflective control.[4] Wittgenstein's and Nietzsche's intuitions differ, however, since Nietzsche is concerned only with consciousness giving way to a neurophysiological materiality, whereas Wittgenstein, in addition to acknowledging that inner episodes such as sensations, feelings, and images might have causes, also believes that explicit thought, reflection, and attention are supported by a repertoire of spontaneous bodily reactions. The bodily repertoire is thus an important frontier where his (experiential) observations, proffered in the service of conceptual analysis, yield to the hypothetical constructions of the psychoneurophysiological sciences that strive to explain life phenomena causally.

A child's body enters into the world with an extremely limited repertoire of "instinctual," or biologically wired, reactions. Reaching for the nipple and smiling in kind are the two reactions most usually cited in this context. The infant's activities are in fact such that it is unclear to what extent they are expressions, that is, to what extent they manifest or signify conditions of life. A newborn does not obviously possess much mind/action (in the sense of the conditions designated in common locutions for mentality and activity).

An important way of describing this situation (and one that links Wittgenstein's late philosophy with that of the *Tractatus*) is that an infant's reactions lack the multiplicity required for the ascription to it of any but the most rudimentary life conditions. The point is nicely made at *LW* I, 942, where Wittgenstein first notes that a newborn cannot be malicious, friendly, or thankful because these conditions are only possible given complicated patterns of behavior. He then adds that a figure containing three straight lines can be neither a regular nor an irregular hexagon. The moral seems to be that the infant's small repertoire of reactions lacks the multiplicity needed to support ascriptions (or gainsayings) of these conditions.

Once a child's repertoire of doings sufficiently expands to exhibit the multiplicity supportive of further ascriptions of life conditions, the lengthy process of teaching and education (*Erziehung*) in acting and speaking has commenced.[5] As any reader of Wittgenstein knows, this process involves extensive assertion, correction, ostensive definition, reference to paradigms, signs of disapproval, citation of rules, observation, participation, attempts, and so on. The result of this teaching, along with the extensive learning that transpires simply through the child's exposure to human activity and its products, is an unceasingly expanding, and increasingly subtler and suppler, bodily repertoire, which now includes verbal reactions. An inner life of sensations and feelings also begins to develop. That the expanded repertoire and intermittent inner episodes also express a vastly enlarged set of life conditions lies partly in the increased multiplicity of bodily doings, sensations, and feelings; partly in the occurrence of doings and inner episodes that are characteristic of an increasing number of mental conditions (as Wittgenstein puts it, "bit by bit daily life becomes such that there is place for hope in it" [Z, 469]); and especially in the steady growth of bodily sayings, in particular, first-person expressive uses of locutions for mentality and activity. In little time the child also learns to offer reasons for its reactions, thus to make first-person explanatory uses of these locutions, and beyond this also to employ these words to report and describe how things stand and are going for her. Words and images also begin to come before her mind.

It is important to stress the contextuality of these expressions. It is not merely the quantitative increase in qualitatively different doings, sayings, and inner phenomena – the "play of expressions," as Wittgenstein puts it – that secures the expression of a wider range of life conditions, but also the enlarged number and expanding sophistication of the circumstances in which these phenomena occur. As the child's knowledge expands, as it undergoes new experiences and acquires firsthand familiarity with wider states of affairs, as its circle of friends, acquaintances, and interlocutors enlarges and it takes an interest in ever more activities and affairs, it acts within an enlarged set of contexts (settings, situations, and practices). It is the embedment of its quantitatively expanded behaviors, feelings, and images within specific configurations of its multiplying contexts that establishes that these phenomena express a growing number of life conditions. This situation is reflected in the following fact. As indicated in the previous chapter, patterns of doings, sayings, and circumstances (along with pictures of the inner) are associated with some condition-of-life concepts. A pattern of this sort, however, is not a visible regularity in bodily performance or behavioral style. It is a set of doings and sayings *in the circumstances in which they occur* that are related by family resemblances. Behavior, in other words, is essentially behavior-in-particular-circumstances (*RPP* I, 314; *RPP* II, 148).

Wittgenstein occasionally describes particular reactions as "primi-

tive" or "natural." These terms are potentially misleading because it
is easy to think of "primitive" reactions as ones a child brings into the
world, whereas in fact a newborn lacks almost all of them. Sometimes
when Wittgenstein refers to a reaction as "primitive" he means that it
is prelinguistic (e.g., *Z*, 541). An example is nursing an injured part
of the body, or attending to the injury of another. Most primitive
reactions of this sort are learned, however; practically all the reactions displayed by, say, a still speechless fifteen-month-old are acquired in familial and other social settings under the impress of
actions performed toward it and through experience of others' activities generally. Occasionally Wittgenstein also writes that a word can
be a primitive reaction (e.g., *PI*, p. 218; *LW* I, 134), a claim that
distances primitive reactions from both the newborn human and
"something animal" (*OC*, 359). What is going on, I believe, is that for
Wittgenstein a reaction is primitive when it underlies a particular
language-game, whether it be innate or learned, nonverbal or verbal.
Before language is acquired, these reactions are necessarily prelinguistic. Once a child has begun to speak, however, verbal reactions,
too, can serve as the conditions of and starting points for mastering
more complex and sophisticated ways of speaking and acting. Similar
remarks apply to the notion of a natural reaction, except that in
some cases "natural reaction" is a pleonasm since "natural" translates
selbstverständlich and thus simply reiterates that the term designates
how someone reacts as a matter of course and without thought in a
particular situation.

    The idea of a primitive reaction, of reactions that adults fasten
upon in teaching children (and others) to speak and act, is central to
Wittgenstein's remarks on language acquisition. Undoubtedly his
most famous example of such a reaction is a child's crying and holding of its hurt bodily part, in response to which adults talk to and act
toward it in certain ways, thereby sooner or later bringing the child
to augment its behavioral expressions of being in pain with verbal
expressions that utilize the word "pain." Contrary to what this and
other examples suggest, however, a primitive reaction need not be a
hook upon which only one language-game can be hung. Although
Wittgenstein never says this, some reactions appear to serve as starting points for the mastery of multiple ways of speaking and acting.
Turning an object over in one's hands, Wittgenstein writes, is the
primitive reaction underlying doubt (*RPP* II, 345); but it also seems
to help underlie other conditions such as curiosity, examination, and
scrutiny. Furthermore, even though a child normally learns many
ways of acting and speaking by performing primitive reactions to
which adults respond in particular ways, it is too strong to claim that
these reactions are necessary conditions for acquiring the doings and

sayings concerned. A child can learn a word such as "think," for example, simply by being exposed to adult usage (see *RPP* II, 241).[6] In some cases, however, reactions are required. For example, it is only to those who react to music in a certain way that the meaning of "expressive playing" can be taught (Z, 164). Moreover, it is important to stress that primitive reactions do not underlie all ways of speaking and acting. The doings and sayings performed, for example, in more rarified and sophisticated domains of life, say, physics, build upon, inter alia, a person's previous mastery of certain concepts and prior acquisition of specific knowledges.

These examples demonstrate, further, that the development of an expressive body depends on the existence of other people who react to it *qua* expressive body. This is an important aspect of the fact that expressive bodies are social products. In treating the child's behavior as expressions of life conditions, people act toward and speak to it in ways that help constitute contexts in which the child latches onto new doings and sayings. Adults do not act in these ways toward inanimate objects and events (and most animals other than pets). Remember, however, that the thesis that expressive bodies are produced by being treated as such, like the story developed in this section generally, does not claim necessity but only that this is how it is in actual human life (to date).

Reactions play a crucial role in the education process both by serving as starting points for the acquisition of ways of speaking and acting and by always underlying and making reflective action possible in even the most sophisticated contexts, for example, international relations and particle physics. (Only people who can spontaneously, say, scan a chart for figures or correct an assertion that Iran is the largest Moslem nation can properly debate foreign policy.) It is important to stress, however, that reactions have a dimension of brute factuality about them. It is simply a fact that throughout the process of maturation some people can and do react in certain ways and other people cannot and do not do so. Wittgenstein calls facts such as this "natural":

> If we teach a human being such-and-such a technique by means of exam-ples, – that he then proceeds like *this* and not like *that* in a particular new case, or that in this case gets stuck, and thus that this and not that is the "natural" continuation for him: that of itself is an extremely important fact of nature [*Naturfaktum*]. (Z, 355)

In designating such facts "natural," Wittgenstein indicates that this is just how things are; there are no explanations for these things by reference to reasons and grounds.

The factuality of reactions is vividly displayed when Wittgenstein contrasts certainty about the sincerity of expressions of feelings with certainty in mathematics. It is a fact about human beings that, having mastered the web of locutions for both feelings and actions connected with feelings, they do not agree in judging the sincerity of expressions of feeling. They spontaneously evaluate one another's expressions contradictorily, and this divergence is a natural fact for which there are no reasons. (These disagreements often also cannot be overcome through the giving of reasons.) Because of this, Wittgenstein writes, uncertainty about the sincerity of feeling-expressions is built into the language-games with the concepts concerned (e.g., *RPP* II, 684). In mathematics, by contrast, people do not, once they have learned to calculate, disagree about the results of calculations. Again, this is simply a natural fact about how people proceed when doing mathematics, and it underlies the type of "mathematical" certainty that is part of this practice (e.g., *PI*, p. 225). If this fact were otherwise, if people did not spontaneously agree in their calculations, this type of certainty also would not exist.

Saying that reactions possess a dimension of brute factuality does not mean that there are no causal explanations of these facts. Against this speaks the widespread mischaracterization of Wittgenstein's remarks as antiscientific in spirit. This topic cannot be examined in detail here. All I will point out is that Wittgenstein's repeated declarations of the irrelevancy of explanations, just like his reiterated exhortations to leave them aside, apply only to certain types of investigation. Examples are those enterprises whose goal is the clarification of concepts, thus philosophy and those areas of the human sciences beset with conceptual problems (psychology, for instance, *PI*, p. 232; *RPP* I, 1063), and also aesthetics along with the interpretive nooks of the social sciences.[7] In the case of philosophy, for instance, the tools of science – causal explanation, hypothesis, and theory – not only fail to help clarify concepts, but only sow confusion when introduced into the attempt to do so. Wittgenstein, however, does not oppose the search for explanations and theories in other areas, including physiology and psychology. One does find occasional statements questioning a mechanistic conception of life (e.g., *Z*, 614), but these are far outweighed by the occasions when Wittgenstein speaks of the causes of both life conditions and the phenomena expressing them (e.g., *PI*, 217, 325, 476, pp. 185, 193, 216; *Z*, 437, 509 [!]; *OC*, 429). He also often refers in a matter-of-fact and uncritical manner to psychological and physiological experiments, explanations, and causation (e.g., *PI*, 412, 493, pp. 203, 212). Of course, Wittgenstein does question whether these explanatory efforts can achieve systematicness. Not only does presumably any type of life condition have a large variety of causes,[8] but there may not be physiological regularities corresponding to certain conditions (*Z*, 608–613). These points, however, proscribe only extremely oversimplifying or overhastily physicalistic theories and can be accepted by most current psychoneurophysiological[9] research.

Acknowledging psychoneurophysiological explanations of life conditions and their expressions does not reduce mind/activity to neurophysiology. Mind/activity comprises a collection of life conditions that are expressed by bodily doings, sayings, images, and sensations. The body's neurophysiology sustains the existence of expressive bodies by housing the causality responsible for all such expressions. But the conditions that these behaviors and sensations express do not thereby reduce to the physical states causing them. Behaviors and sensations do not express conditions by virtue of being their effects. In ways still to be discussed, their doing so is a social matter instituted within social practices. Our physical being simply causally brings about the phenomena that express socially instituted conditions, and thereby maintains and makes possible the existence of mind/action.

To return to the narrative of maturation, the educational process continues for many years, the increasing multiplicity of behaviors and inner episodes in diversifying contexts constituting increasingly sophisticated conditions such as understanding Schrödinger's equations, hoping that one's beloved will return safely from her expedition, and being alienated from the world of parents and everything for which they stand. All along the way, people exhibit differences in what they (can) do and say and thereby in their conditions. Some people also drop out of the steady expansion of expressions and conditions at earlier stages. To the extent that behavior ceases multiplying and assumes tortured forms, we label the individuals involved "mentally" deficient or ill and seek explanations for their distortions and/or inabilities to exhibit certain reactions.

In most cases, however, a person eventually becomes "one of us." This means, first of all, that he or she reacts verbally as we do (e.g., attributes hope without further ado to people who exhibit behavior characteristic of hope). Wittgenstein's way of putting this is that the person masters the technique of language, where "technique" means customary and regularized way of using or, more perspicuously, customary and regularized way of behaving (*Behandlungsweise, RPP* I, 1025). Mastering a technique is joining in with customary and regularized ways of reacting with language. Once the sorts of reason standardly employed to justify reactions are exhausted (e.g., she said such and such, that's why I said she hoped he'd come), one is left with the brute, socially conditioned, and neurophysically underpinned fact at the heart of every technique that language just is used like this. ("Why does her saying this indicate that she hoped he would come?" – That's what we say of someone who says things like that in such circumstances.) Mastering the technique of language also means of course that a person uses language as others do in reflective and considered speech.

Being one of us means, more broadly, that a person speaks and behaves intelligibly to us. Vis-à-vis speech acts this means, among other things, that what a person says and writes when confronted with novel and unusual situations is immediately understandable to the rest of us. It also means that when someone departs from familiar usage, he can make himself understood relatively easily with a succinct explanation of the departure. Suppose, for example, that someone ascribes desires to a person different from those that we would ascribe to her given her strange, repetitive behavior. He still counts as one of us if he can make himself intelligible through a concise explanation of his theory that frustrated sexual desire sublimates into such behavior.[10] For a concise explanation can succeed only if his theory does not institute new techniques for most of our life condition locutions, that is, only if it largely perpetuates existent techniques for such locutions. (Being one of us in this wider sense thus requires being one of us in the narrower sense.)

Vis-à-vis doings, being one of us implies simply that someone's actions are intelligible. This means, first, that her actions make sense to us in the contexts in which they occur and, second, that we grasp which life conditions they express – for example, that a dismissive gesture manifests frustration, that a shaking of the head signifies belief that the rope will not hold, and that a mad dash constitutes rushing to bus stop. When these things are not understood without further ado, a person still counts as one of us so long as her behavior becomes intelligible once we learn of reasons for her action or about the contexts in which she acts.

Since human doings- and sayings-in-particular-circumstances are immensely complex, the frontiers of who we are are rarely firmly delimited. One never fully understands anyone, even those closest to oneself; and just where someone begins to become unintelligible and thus no longer one of us is a contingent and shifting matter. Given, furthermore, commonalities in more or less all human lives – in physical environs, such facts of life as birth, death, and sex, basic needs and emotions, such practices as teaching and celebrating, and some primitive reactions – practically all human beings are intelligible to us at least to some extent; and just when it makes sense to say that we finally understand someone from a distinctly different culture is again a contingent and fluctuating matter that often depends on her attitudes toward us and not our powers alone. A we's internal boundaries are similarly flexible and porous, since the border between sanity and insanity, between those who behave intelligibly and those who, despite the same education, do not act and speak like us, is subtle and shifting. Foreignness, sanity, and we-ness are all matters of degree. Despite this, we's can be carved out of the complicated mosaic of human existences, for it is possible roughly to identify open-ended sets of people who use language alike and act and speak intelligibly to one another.

The people who are "us" are the people with whom "we can find our feet" (*sich befinden in, PI*, p. 223), more literally, the people in whom we can find ourselves. They are those with whom we share a kind of form of life. Some Wittgenstein commentators have highlighted the notion of a form of life, believing that it sums up crucial aspects of his philosophy and is a key to unlocking it. I agree with those who claim that the term is vague and a poor mine for illumination.[11] Not only will whatever detailed understanding can be had of the term derive from our understanding of his ideas generally, but the specific contexts in which it is used do not decide between certain interpretations of it. This is the situation, it seems to me, because Wittgenstein uses the term in a colloquial fashion to mean something like a way of living, or a way life flows on among people. People share a form of life in sharing a way life flows on. The scope of the term cannot be confined, however, to social, cultural, or species instances. What I mean is that forms of life are not inherently aspects of either cultures, societies, or species alone. They can be any and all of these. How life flows on can be how it flows on in a particular culture, society, or, moreover, community, discipline (e.g., mathematics), or family (etc.). In such cases, a form of life is an open-ended set of contextualized ways and patterns of acting and speaking, which initiates learn to carry forward and which helps constitute the sociocultural entity in question. How life flows on can also be, however, how it flows on among humans, in which case the form of life involved is that of a species and consists in ways of speaking and acting common to humans along with such facts as that humans can speak and that they react like others after receiving a certain type of education. When, consequently, Wittgenstein writes that "to imagine a language is to imagine a form of life" (*PI*, 19), he claims only that to imagine particular speech acts is to imagine a way life flows on of which these acts are part. Saying this does not itself further qualify the flow socially, culturally, or "speciely." To the question, then, How many forms of life are there?, the answer is, It depends – many if you've got sociocultural ones in mind and one if you are asking about humans per se.

That commentators have discovered pluralizing sociocultural and singularizing biological connotations in the term[12] and, on that basis, argued that a form of life is essentially a sociocultural matrix or biological entity reflects the enmeshing of society and biology in ways of speaking and acting. The social dimension consists at once in the fact that ways of acting and speaking are practices that embrace multiple individuals, and in the embedment of the actions that constitute these practices in social contexts. (Sociality is also of course reflected in differences among the practices comprising different forms of life.) The biological component consists in limited commonalities in basic ways of acting and speaking; in the human capacities to speak and to respond to the sort of education received within forms of life; and in the contribution made by genetically determined bodily structure and function to the brute factuality of the situation, that people reared within particular forms of life react like this and not like that in particular circumstances.[13] Thus, in sum, although "form of life," like "language-game" (*PI*, 23), crystallizes Wittgenstein's general attitude toward language, it does not designate any one sort of well-defined entity and should be used gingerly in philosophical monographs or accounts of human existence inspired by his thought.

Becoming one of us, participating in our ways of speaking and acting (sociocultural forms of life), is as far as most human beings develop. Some individuals, however, are capable of extending the bounds of the intelligible in recognizably novel ways. *Qua* novel, their doings and sayings (and products) often, though not always, share with what is foreign or insane the veneer of unintelligibility. When this occurs, these doings, sayings, and products nonetheless differ in emanating intimations of insight and originality and in being performed by people who otherwise largely speak and act intelligibly. Of course, it is only from the vantage point afforded by participation in a sociocultural form of life that novelty appears as such and thereby differentiates itself from the foreign and insane. The way Pablo Picasso's multiperspectival images of faces and people were novel to Westerners in the 1910s differs considerably from how they would have been foreign to a !Kung bushman had he stumbled upon them. In any case, individuals who through their acts extend the bounds of intelligibility are beyond technique: They do not simply go on using language as we do and behaving intelligibly to us, but transcend these in ways that eventually draw us along, expanding the horizon of what we understand, transforming how we speak and act. Of course, even technique has its own form of novelty, consisting in the extension of existing ways of speaking and acting under new and ever changing circumstances. But although the evolution in speaking and acting that arises therefrom is omnipresent,[14] it is usually almost invisible in comparison with those modifications induced by the rarer, but conspicuous novelties that startle and confront established ways of going on.

I will illustrate the phenomenon of novelty by describing three sorts of person who extend the limits of the intelligible in different ways. (This list is not meant to be exhaustive.) The first is the visionary who makes people see things afresh in offering large-scale perspectives at odds with reigning ones. Examples are Jesus Christ, Galileo Galilei, René Descartes, the Buddha, Charles Darwin, and Sigmund Freud. Their novel perspectives are not confined to but instead transcend particular disciplines and walks of life.

A second such type embraces people who create objects of what Wittgenstein calls at one point "intransitive" understanding. Intransitive understanding is a grasp of perceptual objects that helps constitute (some) experiences of the objects and cannot be adequately expressed in words. The paradigm objects of such understanding are for Wittgenstein musical compositions, but works of art, poems, and dance pieces are further examples. Understanding these objects is akin to an experience that can only be indirectly conveyed – by suggestions and hints about how to hear, view, or read them, comparisons with other such objects, metaphors borrowed from other sectors of language, and expressive gestures.[15] Confronting such objects extends intelligibility if it occasions understanding that transcends or transforms extant ways of speaking and acting (this need not occur). Certain usually illustrious composers, artists, choreographers, and poets fall into this second category.

The third group comprises experts in understanding human beings, the people who repeatedly make incisive judgments about how things stand and are going for others (see *PI*, p. 227). They extend comprehension of human life by offering insight into what makes people tick on particular occasions or in general, thus enabling us to use our concepts for mentality and activity with greater suppleness and lucidity. I am tempted to label people in this category "wise," for I want to include within it experts in moral judgment;[16] and wise is paradigmatically he or she who can repeatedly judge people and their moral situations insightfully, decisively, and perspicuously. In this category one finds quite a hodgepodge of people, many of them of no public or academic repute.

Most human beings, however, simply join in with extant sociocultural forms of life, and some never succeed even at this. Although practically all humans begin their lives with the same extremely limited behavioral repertoire, development scatters them along a spectrum of expressive multiplicities ranging from the less to the more diverse, flexible, and inventive.

The Institution of Life Conditions

The preceding sketch of the progression of a human being from infancy onward shows that a functional adult's extensive bodily repertoire of doings and saying is social in the sense of being acquired through learning and training in the context of others' activities. (This is also true of sensations, feelings, and images since a person's undergoing these is tied to linguistified awareness, and language is mastered within social practices.) An expressive body that manifests, signifies, and constitutes conditions of life is thus a social product. Production of expressive bodies is not, however, the only way mind/action is socially constituted. Indeed, little was said in the previous section about conditions of life beyond the obvious point, that as a human being matures it expresses increasing numbers of and increasingly sophisticated conditions. I want now to examine a second dimension of the social constitution of mind/action by considering four features of mind/action that characterize people *by virtue of* their participation in social practices, their maturation within a context of people reacting, speaking, correcting, and continuing in certain ways.

The first such feature is being in conditions of life. To be in a condition of life is for things to stand or to be going some way that is expressed in doings, sayings, sensations, and images. This expressive relation is almost entirely socially constituted. For: That condition A is what is expressed by such and such a behavior or inner episode performed or transpiring in circumstances C, and that therefore one of us is A when performing that behavior or undergoing that inner episode, is a matter of the behavior or inner episode concerned, the circumstances C, and understandings of concepts of life conditions.[17] Since for Wittgenstein understanding the concept of X is understanding what X is, I will henceforth speak alternatively of understanding concepts of life conditions and understanding conditions (i.e., what they are). One advantage of the latter form of expression is that understanding a particular life condition, for example, hoping he will come (versus hope *simpliciter*), is not usually described as understanding the concept of that particular condition. Yet understanding particular conditions, as much as understanding types thereof, is part of the background against which behavior and inner episodes express particular conditions. I emphasize, furthermore, that on Wittgenstein's account, understanding (the concept of) X is expressed not only in uses and explanations of the linguistic expression "x" (as well as occasionally in uses of other expressions), but also in reactions to the phenomena or persons qualified as x (e.g., Z, 543; *RPP* I, 910).

To grasp better the thesis of social constitution, consider first a case where the expressive relation is biologically determined. As sketched in the previous chapter, some wired-in expressions of simple conditions (e.g., crying and being in pain) are augmented through social education by further expressions, including verbal ones. Because these further expressions augment earlier ones, their performance in certain circumstances manifests the condition expressed by the prior ones. But not only, in addition, is the crying that expresses being in pain biologically wired. That being in pain is what crying expresses is a biological fact. That is, the connection between the behavior and the condition, namely, that *this* condition is what *this* behavior expresses is biologically established. Crying, it can be said, is the biological expression of being in pain.[18]

Of course, the concept of pain used in these sentences is a complex one expressed in elaborate ways of speaking and reacting. The propriety of using it in formulating what wired-in behavior expresses rests on its having taken over the biologically established connection between crying and being in pain. Evidence for this incorporation is the continuity in reactivity that reaches from primitive reactions to crying (e.g., attending to the cryer) to the more elaborate ways of reacting and speaking that express the full-blooded, developed concept. At, consequently, the core of the extended multidimensional concepts of pain found in natural languages lies understanding that crying expresses this condition. And this later fact is a matter of human biology. So an adult who ascribes pain to an infant on the basis of its crying does so dependent on his complex and mature understanding of pain. But that on this understanding crying-in-its-circumstances expresses pain has a biological source. It might be added that the developed understandings of pain possessed by people living at different times and places will thus more than likely share a common denominator, however much these understanding transcend this commonality in concordance with the greater complexity of mental life implemented in social existence.

Most conditions of life, however, have no biologically established expressions. Contentment, expressed in smiling, might be one of the few exceptions. The understandings, accordingly, given which behaviors-in-circumstances express conditions of life have at best a thin biological fundament. That given behaviors express specific life conditions is for the most part socially established.

Recall that many condition of life concepts, in particular those for emotions and states of consciousness, are associated with life patterns, collections of behavior-in-circumstances related by family resemblances. This association partly reflects the fact that which (if any) such conditions are expressed by particular behaviors-in-circumstances depends on the behaviors involved, the contexts in which they occur, and understandings of life conditions. For the (relative) continuity of these understandings across time ensures that the different behaviors-in-circumstances that express a specific condition form a pattern. (Conversely, the reappearance of behaviors-in-circumstances that compose patterns helps secure the continuity of such understandings.) These understandings, as just noted, rarely have biological roots. The association of patterns and conditions is socially instituted, consequently, because it rests upon developed complex (conceptual) understandings of hoping, fearing, imagining, and listening carefully that are expressed in elaborate ways of speaking and reacting to the phenomena involved. (Remember, moreover, that the body that performs these behaviors, thus the behaviors too, are social products.) That specific behaviors are characteristic of any such condition is likewise tied to the understanding of that condition. Of course, nothing is sacrosanct about our present understandings of life conditions. Conceivably facts will be discovered about the phenomena that express conditions that so change what we understand by hope, fear, and so on (even pain) that the old patterns no longer are associated with these conditions. (This can also occur via transformations in reactivity.) However, because (1) our present understandings are what would initially inform us that what had been discovered were facts about expressions of these conditions, and (2) conceptual understandings of life conditions are broadly solidified in social time and space, we should not expect the connection to break down anytime in the foreseeable future.

Not all conditions of life exhibit characteristic behaviors or are associated with patterns of behavior-in-circumstances. Moreover, a person can be in a particular emotion or state of consciousness even when his behavior is not characteristic of that condition or part of the pattern associated with it. Performing an action, holding a cognitive attitude, and being in a mental condition on these occasions is still, however, a social affair. Given a play of doings, sayings, sensations, and images, a person is performing a specific action and/or in a particular cognitive condition or uncharacteristic mental one by virtue of the expressions, their contexts, and complex understandings of the conditions involved. For, once again, *given these understandings,* behaving this way (etc.) in these contexts signifies that one is in the condition involved. Moreover, being in a cognitive condition that goes unexpressed, and whose duration is marked only by the truth of the counterfactual conditional(s) that had particular circumstances occurred the person would have acted in a certain way, similarly rests on practices of speaking and acting which establish that the holding of the counterfactual signifies the condition involved.[19]

This formulation slightly oversimplifies matters. Strictly speaking, understanding determinately institutes specific conditions only in conjunction with the entirety of the contexts in which behaviors and inner episodes occur. These contexts, furthermore, are invariably complicated, embracing events in the immediate settings of action, wider social situations and practices, past and future behaviors and inner episodes, other conditions of life, and physical states of bodies. What is more, even given specific expressions and the totality of contexts in which they occur, conceptual understanding might not unambiguously determine that a person is in some specific condition. Common locutions for mentality and activity are extremely flexible, and different words can often equally well capture how things stand or are going for someone at some point. For example, the totality of circumstances might exclude someone's hating, thinking of, expecting, and seizing (etc.) him, while leaving open whether desiring or wanting is the proper specification of what she is about. Think also of the indeterminacy that occasionally affixes the choice among "understanding," "comprehension," "cognizance," "apprehension," "grasp," and "knowledge."

That the contents of mind/action can be indeterminate and ambiguous should trouble no one. How things stand and are going is *essentially* articulated in what people do and say. It is not an already formed thing that they try to describe in words.[20] Accordingly, mentality and activity contents share the flexibility and ambiguity characteristic of the techniques for life condition locutions. Indeed, this tie to language underlies the use of metaphors and poetic language in this context. In becoming one of us, a person learns how to use language to articulate how things stand and are going for her. At first this ability is confined to a few simple words for conditions. As, however, the complexity of life increases – the contexts relevant to existence multiply, the repertoire of doings and sayings expands, the intermittent procession of inner episodes enriches, experiences of behaviors and others' self-descriptions gather, and a larger arsenal of life condition concepts is mastered – she discovers the flexibility of the language of human expression and that what is conveyed by words for mentality/activity can sometimes be more poignantly and penetratingly disclosed through metaphors and poetic sayings. Common words for mentality and activity are simply age-old, handed-down, and endlessly repeated expressions with which people articulate how things stand and are going. Metaphors like "it burns like fire" are also widely harnessed, taught, and passed down as common locutions. The gifted, meanwhile, avail themselves of more delicate and nimble poetic verse as a more acute and incisive vehicle for expressing these matters.

The intuition that mental contents are unequivocal and determinate, along with the theory that they are so either each taken for itself, or by virtue of their interrelations, or by virtue of their interrelations together with their relations to behavior, the physical environment, and/or even social context – this intuition and theory are descendants of the picture of mind as a thing, realm, or stratum with distinct states or occupants. I do not claim that they presuppose or rely on this picture. Indeed, the linguistic approach to mental matters pervasively adopted in recent decades often prescinds from ontological assumptions and implications. I maintain only that the compellingness of the intuition (and theory) reflects a molding of sensibilities and habits of thought under the impress of this picture and on the background of its historical sway. Once the picture is abandoned, and once it is seen that forsaking it does not entail denying that people have sensations, images, and words "before their minds," the idea that mind/action is connected essentially to conceptual understandings and is not what it is independently of these understandings becomes more plausible. That its contents are occasionally indeterminate and ambiguous therewith also makes more sense.

In addition, in the rough and tumble of real life we are of course rarely apprised of all contexts of behavior. In judging others' conditions, consequently, we rely on "criteria," as Wittgenstein puts it: behavioral phenomena by virtue of which it makes sense to say – until we learn of further aspects of the contexts that defeat this – that someone is in a given condition. People's evaluations of the bearing of specific contexts also often deviate. So our judgments of others' conditions diverge, and we occasionally challenge a person's self-descriptions. Insofar as agreement can be restored in these matters, it is achieved either through further elucidation of context or through a discussion of what believing, fearing, expecting, and the like are.

Someone might reply at this point, "But it is quite remarkable that a person ever knows the right behavior to perform, that is, the behavior that expresses her condition. She is, after all, in the condition she is in. Yet you claim that it is her behavior, sensations, and images, together with circumstances and conceptual understandings, that 'constitute' her condition. So it seems that she must be familiar with these understandings, know her circumstances, and quickly perform no mean calculation in order to be able to perform the right behavior." Obviously no such process occurs. An angry orator who believes that such and such must be done does not usually need to consider what behaviors properly express these conditions. She just reacts and, in doing so, expresses her conditions. She is of course familiar with life conditions and her situation, but this familiarity need not be explicitly drawn upon. She just acts. And what conditions are thereby expressed depends on what she does and says, the contexts in which she acts, understandings of life conditions, and possibly also whatever inner episodes (if any) occurred. So yes, she is in the conditions she is in. But *which* ones these are (e.g., anger and particular beliefs) depends on behavior, circumstances, and understanding. It isn't what it is independently of these matters.

"Fine," the questioner might reply, "my point wasn't really that she must explicitly consider the matter. Rather, doesn't she perform the behavior in question *because* she is, say, angry? And isn't this what qualifies her behavior as an expression of anger, not any business about circumstances and concepts?" Yes, she acts that way because she is angry. But the "because" involved, as many writers have argued,[21] is the because of the elucidation of intelligibility, not the because of causality as the questioner's formulation hints (cf. *RPP* I, 217). When we say that she acted thusly because she was angry, we mean that anger makes sense of, offers a reason for, her behavior in its circumstances. (Recall, as discussed in Chapter 2, that conditions of life must meet additional provisos to afford intelligibility explanations.) But not only is the because not that of causality. Saying that she acted because she was angry *presumes* that her behavior expressed anger. It doesn't yet indicate what qualifies it as such an expression. Suppose that she was under pressure at work, that things were going badly with her girlfriend, that she had hoped for support from her colleagues, and that she has a history of snippety interactions with that other person. When, consequently, he commented on her report, she reacted as she did. When someone reacts like *that* in *those* circumstances, then given our understanding of life conditions, we say that she is angry; and it is because of this understanding that mentioning anger to someone who witnessed her behavior but does not know her situation and history informs that person why she behaved thusly. So even though she acted as she did because she was angry, what qualifies her behavior as an expression of anger is not its being the effect of anger, but the place it occupies in the play of expressions and contexts of behavior, in conjunction with our understandings of how things can stand and be going for people. Incidentally, if one wants to identify a cause for her behavior, the comment to which she responded will serve as well as anything.[22] Her movements were also presumably brought about by neurophysiological pathways.

So much for the first feature of a person's life consequent upon integration into social practices. The second feature is the abilities to describe, explain, and report one's conditions to others. From Wittgenstein's perspective, not much can be said philosophically on this topic. What marks off conditions of life from the physical states also ascribable to an individual are ascriptional asymmetries, the fact that whereas others ascribe such conditions to someone on the basis of his contextualized behavior, he (usually) ascribes them to himself on the basis of nothing at all. A person, that is, does not go on anything in describing and reporting his own conditions. This holds even when challenges to claims about, say, past beliefs – that one believed yesterday that John was on vacation – are met with counterfactual assertions to the effect that, "If you had asked me then, I would have told you that he was still away." A person (usually) goes on nothing in making such assertions. He simply utters what he cannot doubt.

"Nothing" means that a person does not go on anything, at least anything that he or she can cite as evidence, criteria, marks, intimations, or whatever. Another way of putting this is that there is nothing articulable or designatable in the experience of a person whose understanding is formed within social practices that informs him what his conditions are (remember that this holds for conditions of consciousness too). At the same time, a person's ability to talk about himself presupposes familiarity with circumstances as well as conceptual understanding and to this extent resembles others' ability to talk about him. And, to repeat a familiar point, there might be causal explanations of people's self-explanations, -descriptions, and -reports. In Wittgensteinian language, there might be explanations that reach beneath the (phenomenal) "surface" to explain *the phenomenon* of people telling us about their conditions.

A third feature of mentality/activity consequent upon participation in practices is the ability to identify others' conditions. This ability takes two basic forms. The first is the ability to *perceive* (primarily see and hear) that someone is in such and such condition. This ability pertains mostly, though not exclusively, to mental conditions manifested in behavior. Wittgenstein states emphatically that one does not infer from, say, fear or pain behavior that another person is fearful or in pain; one sees it:

"We *see* emotion." – As opposed to what? – We do not see facial contortions and make inference from them (like a doctor framing a diagnosis) to joy, grief, boredom. We describe a face immediately as sad, radiant, bored, even when we are unable to give any other description of the features. – Grief, one would like to say, is personified in the face. (*Z*, 225)

This also applies to actions.

This kind of seeing obviously differs from the seeing of physical objects. For instance, it demands familiarities not required in the latter case; namely, with human behavior generally, the patterns of behavior-in-circumstances associated with mental locutions, and the

specific contexts of particular behaviors. Wittgenstein refuses to follow those philosophers who insist that the objects of sight are "strictly
speaking" or "in the first place always" the physical attributes of
things. He accepts common usage, on which it makes sense to say
that people see emotions, actions, and conditions of consciousness.

Being able to perceive others' mental conditions rests crucially on
familiarity with the patterns of behavior-in-circumstances associated
with these conditions. For seeing a person's mental condition usually
entails understanding her circumstanced behavior as part of such a
pattern. These patterns, as noted in the previous chapter, are inherently irregular. Learning to recognize one, consequently, is acquiring
senses for (1) the particular flavor of circumstantial relevancies on
the basis of which specific behaviors express the condition in question, and (2) the lines of intelligibility that bind different combinations of behaviors and circumstances into a single pattern. These
two senses centrally constitute a person's understanding of the life
condition concept associated with the pattern, hence his or her understanding of the condition itself. Being able to see that this here
behavior expresses condition A thus requires (a) understanding of
(the concept of) condition A, that is, familiarity with the configurations of relevancy and lines of intelligibility that bind irregularity into
a pattern, and (b) acquaintance with the behavior's specific contexts.
Of course, this same (conceptual) understanding also enables a person more widely to attribute and talk about conditions of life intelligibly.

Talk about patterns has aesthetic overtones. This suggests that the
ability to perceive others' conditions might resemble one or both of
two experiential abilities/understandings that Wittgenstein discusses
in relation to aesthetics broadly conceived: intransitive understanding and seeing-as. I will provide the very briefest characterizations of
each. As described, intransitive understanding is an unformulable
grasp of a particular object, which helps constitute an experience of
the object and whose point is wholly immanent in undergoing the
experience. Objects of such understanding are musical compositions,
poems, artworks, dance pieces, statements (*PI*, 531), and faces (see
PI, 531–539). Seeing-as, meanwhile, is a visual experience that a
person sometimes undergoes when looking at certain objects, paradigmatically certain drawings such as the famous duck-rabbit. It involves seeing the drawing as one thing rather than another, exhibits
a strong component of conceptuality, presupposes a multiplicity of
aspects that the drawing can be seen as, and undergoing it is subject
to the will.[23]

Perceiving someone's mental condition resembles intransitive understanding in that perceivers will often be unable to identify what it

is (if anything) in others' behavior on the basis of which they perceive conditions (e.g., *PI*, p. 228). It differs, however, since it can and is articulated in the form of reports or descriptions of how things stand or are going for others. It further differs from intransitive understanding, and seeing-as too, in not being a "special" or relatively rare experience, but one that regularly occurs in the face of behaviors performed by participants in one's own forms of life. It also differs from seeing-as in not being subject to the will.

The second form taken by the ability to identify others' conditions is the ability to infer them. Conditions ascertained through inference are usually either attitudes or actions and only occasionally emotions or states of consciousness. Just like the ability to perceive conditions, the ability to infer them rests on a grasp of configurations of relevancy and lines of intelligibility. In this case, however, the configurations and lines do not figure patterns. Not only is someone's believing something, hoping for something, knowing something, or intending to do something rarely expressed in behavior that falls into an associated pattern; sometimes it is not expressed in any behavior at all. (This is why it is often impossible to perceive intellectual attitudes and actions in behavior.) In comparison with the configurations and lines that shape perception of mental conditions, those relied upon in inferring cognitive ones are more abstract and less phenomenally insistent, more strongly dependent on knowledge of others' contexts, and more extensively supported by presuppositions and reconstructions.

Another important difference between perceiving and inferring conditions is that inference often rests on a reconstructed web of belief and knowledge that is attributed to someone largely on the basis of her community, biography, positions, and the like. Often, in fact, inference is supported almost entirely by reconstructed webs (and other contexts) and hardly at all by behavior. Inference nonetheless resembles perception because in both cases ascertaining how things stand or are going depends on behavior, contexts, and the understanding of life condition concepts. Conceptual understanding is alive in inference because it comprises the abstract relevancies and intelligibilities that underwrite not just inference, but the reconstruction of belief webs too. (Many of these relevancies and intelligibilities, moreover, are molded by earlier inferences.)

Before considering a fourth and final feature of mind/action consequent upon integration in practices, I would like to explain, though it may be obvious, that the position described here does not succumb to the problems that Jerry Fodor and Charles Chihara famously find in a position they call "logical behaviorism."[24] An important reason for this immunity is my not analyzing understanding (concepts) via criteria (and I do not believe I diverge from Wittgenstein here, though the point will go unargued). Fodor and Chihara define a criterion as follows: A behavior is a criterion for a condition when the meaning of the word for the condition justifies the claim that in normal circumstances (of type C, which they leave unspecified and unanalyzed) one can attribute the condition on the basis of the behavior (p. 397). Note, first, that for Wittgenstein understanding a concept is not the same as understanding a word (a concept is not the meaning of a word). A concept is more like a unit of organization for states of the world that is expressed not only in how people speak, but also in how they act toward the phenomena designated by their words. Formulated in terms of concepts, then, Fodor and Chihara define a behavior as a criterion for a condition when the (understanding of the) concept of that condition justifies the claim that in normal circumstances the condition can be attributed on the basis of the behavior. This definition implies that normally it is conclusively justified for someone who understands the concept to attribute the condition on the basis of the behavior concerned. That this is the case is also something one learns in mastering the concept. Indeed, one learns the concept by being taught that this and that behavior is conclusive evidence for the condition in normal circumstances.

On the analysis presented here, one does not learn criteria in this sense in coming to understand a life condition concept. One instead grasps configurations of relevancy and lines of intelligibility. These configurations and lines are such that, given certain behaviors and what is known about their contexts, it makes sense to attribute the condition. There is no question of conclusive evidence, only of what it makes sense to say, since understanding the concept also comprises the comprehension of indefinitely many *possible* further aspects of the situation, the existence of which would defeat the attribution. The phrase "normal circumstances" gestures in this direction, but in a misleading way. For "normal circumstances" are variable, sometimes so variable that what the phrase designates becomes indefinite and the phrase accordingly loses its point. It is easy to overlook this vis-à-vis such states of consciousness as pain that are vividly manifested in behavior. But it is obvious in the case of emotions, cognitive conditions, and many actions. The situational relevancies grasped in mastering concepts of these conditions, relevancies on the basis of which it makes sense to attribute these conditions, do not reduce to any list of normal circumstances. They permit the identification of such circumstances, but do not reduce to them.

Furthermore, its making sense to attribute a specific condition rests on understanding of the concept of not only that condition, but others as well. In the face of specific behavior-in-contexts, it makes sense to make any one as opposed to a different attribution on the basis of the grasp of life conditions generally. Thus I agree with Fodor and Chihara that in acquiring these understandings a person grasps "complex conceptual connections which interrelate a wide variety of mental states" (p. 413); and that, as a result, children might learn the concept of, say, dreaming not only by being told, when they report strange goings-on in their sleep, that they had dreamed, but also because, "having grasped the notions of *imagining* and *sleep*, they learn what a dream is when they are told that dreaming is something like imagining in your sleep" (p. 415).

It is not possible, in other words, to "operationalize" understandings/concepts of life conditions (or, for that matter, the meanings of words) in terms of well-defined sets of behaviors. Even the patterns associated with mental conditions do not evince well-defined behaviors, since they are family-related sets of behavior-*in-circumstances* and not regularities in behavior. Fodor and Chihara thus "overlook" (cf. pp. 410–412) the possibility of a noncriterial conceptual connection between behavior and conditions, of a noncriterial conceptual justification for attributing conditions on the basis of behaviors-in-circumstances. Contrary to what they write, neither the recognition of plural nor the introduction of new "criteria" for the application of a concept ("criteria" in the sense of empirically hypothesized and supported evidences) amounts ipso facto to fragmenting or changing the concept. When science introduces new criteria in this sense for, say, dreaming, it does not therewith automatically fashion a new concept and change the subject. Most of the criticisms Fodor and Chihara rain upon logical behaviorism in section 6 of their essay thus pass the position in this study by.

Moreover, it is only in an epistemological context that attention is restricted to behavior. When the issue is what a person's conditions are and not how we know this, inner episodes are also relevant. The remainder of the problems that Fodor and Chihara ascribe to Norman Malcolm (logical behaviorism) are therewith avoided. For instance, it is unproblematic to suppose that people can dream, systematically forget their dreams, and never therefore verbally express them. For, when sleeping, they might have something analogous to the sensations, images, and the like had when awake. Moreover, science might discover physical evidence for these dreams (e.g., causal pathways causing inner episodes). Of course, none of this implies that Fodor and Chihara would accept the account outlined here, but that is a different matter.

A fourth – and final – feature of mind/action consequent upon integration into social practices is the possession of "convictions" that, in Wittgenstein's words, "hold fast" in a human life. Although this topic has been extensively discussed in the literature, it is worth providing a summary here. Learning to speak and act in particular ways involves accepting indefinitely many "convictions," whose correctness is not ascertained and of which it is not the case that a person holds them because he is satisfied of their correctness (even though he is nonetheless in no doubt about them). Examples of such "convictions" are that my body has never disappeared and reappeared after an interval, that the earth exists, that I've never been in the Andromeda Galaxy, and that two plus two are four. "Conviction" is not really the right word here because convictions usually concern explicitly considered matters, and what comes to stand fast for someone through learning to act and speak in certain ways usually *goes without saying*. What stands fast is not for the most part taught, and it need not ever be formulated (see *OC*, 94–95). Because it comes to stand fast for someone through his learning to do something, a person also can only "discover" that his believing, arguing, reflecting, and the like rotate around it (*OC*, 152). There is, furthermore, no particular evidence for these obvious matters, since any putative evidence couldn't be more certain than they themselves. Wittgenstein writes, accordingly, that if events were nonetheless to occur that placed one or more of these matters in doubt, then other obvious things, judgments generally, and one's grasp of linguistic meanings would all be thrown into question.

What stands fast for a person is indefinitely complex. As suggested, moreover, something's standing fast does not require that the person for whom it does so ever entertain it. Not only is it impossible to consider everything obvious, but explicit entertainment is not needed in order for matters either to be firm or to serve as background for behavior. The settledness of these matters consists in ways of acting and speaking being such that if these matters ever *are* formulated, they are exempt from doubt (see *OC*, 88). Their standing as background, meanwhile, consists in the fact that if doubt is successfully cast upon them once they are formulated, not only does much of what stands fast tumble, but a person's grasp of meaning and concepts is also shaken. Instead of describing background obviousnesses as forming "webs," consequently, it is better to speak of them as serially and indefinitely articulable on the basis of people's understandings of how to act and speak. Through the acquisition of these understandings within social practices, much automatically comes to stand fast and be obvious for people. It is unnecessary, consequently, to suppose that what stands fast is in any sense stored in people. The use of the expressions "to believe," "to know," and "to doubt" vis-à-vis what is obvious does not support the supposition of either a mental storehouse (e.g., the aviary in Plato's *Theaetetus*) or a filing system in the nervous system. Rather, people's understanding of the concepts of belief and doubt, in conjunction with their wider ability to grasp particular languages, is simply such that when certain statements are made, perhaps for the first time ever, people are not in the position to doubt them. (These abilities also enable people to produce such statements indefinitely upon demand.)

Wittgenstein wrestled with the question of whether this background "supports" acting and speaking. Sometimes he writes that the background is the element in which argumentation, reflection, consideration, weighing of evidence, and the like "are alive" (*das Lebenselement der Argumente; OC*, 105, and cf. 167, 411). At other times he suggests that the background is "indubitable" only by virtue of the remainder of people's doings and sayings (e.g., *OC*, 103, 153, 210, 248). No consistent answer emerges from his remarks on this question, and perhaps the two types of claim are not really at odds. In any event, the background comprising what holds fast must be distinguished from the webs of presupposition that underlie investigations in specific arenas (see, e.g., *OC*, 402). This is so in part because Wittgenstein withholds the appellation "presupposition" from the background on the grounds that presupposition implies doubt (*PI*, pp. 179–180). (He did not, however, seem to reach a settled conviction about whether it is right to call what stands fast "assumptions.") Particular investigations, discussions, and examinations thus possess a double background: a specific set of presuppositions that forms the starting point for this particular work, and an indefinitely articulable general background of what stands fast for speaking and acting more widely.

To close this section, let me stress the reciprocity that links conditions of life, avowing and identifying such conditions, and holding background convictions. To begin with, these four features interdependently characterize an individual's life. A person's conditions, for instance, are interwoven with her self-ascriptions, her perceptions and descriptions of others, her background convictions, and the bulk of her other doings and sayings (and inner episodes). Second, and of perhaps greater significance, that these features interdependently characterize one individual is tied up with their interdependently characterizing others. These features characterize an individual by virtue of her participating in practices that encompass open-ended numbers of people who act toward and talk about one another in mutually intelligible ways. As we have seen, moreover, one person's conditions generally depend on how others act toward and talk about her, one another, and themselves. How they thus talk and act itself rests upon both their conditions and those they identify in others. So one person's conditions depend on others' conditions, those they all recognize in still others, and the ways they talk about anyone. One can go round and round with formulations like this. The main points are, first, that any individual's being, doing, and understanding are not only interwoven with but, to varying extents, the same as others'; and, second, that this interwovenness and commonality arise from the incorporation of individuals into practices, which establish lines of intelligibility and configurations of relevancy along which lives consequently flow.

The Constitution of the Individual

The first section of this chapter discussed how expressive bodies are produced through the enclosure of human bodies within social practices. Exposed to the doings and sayings constituting these practices, and treated as an expressive body in teaching, correctional, and other actions taken toward her, a person learns to perform myriad behaviors that perpetuate existing behavioral patterns, discourse about the world, and are highly sensitive to a tangled network of mostly social contexts. Her behavior thereby comes to express, for both herself and other participants in the same practices, a large variety of life conditions.[25] The second section examined how social practices, in addition to producing expressive bodies, institute mind/action. Given the (conceptual) understandings carried in such practices, people who perform certain behaviors and undergo particular inner episodes in particular circumstances are in such and such conditions.

I wrote in the previous chapter that a body's activity, in expressing conditions of life, at the same time establishes that there is an individual who is in these conditions. It follows that because social practices fashion expressive bodies and institute conditions of life, they also constitute individuals, namely, the people who are in and to whom can be attributed the instituted conditions expressed by fashioned bodies. The social constitution of the individual does not, however, amount simply to the formation of the expressive body and the institution of mind/action. Being an individual is a multifaceted affair. In addition to possessing mind/action, a person, inter alia, has an identity(ies), displays a character, exemplifies person types, and exhibits subjecthood as well as gender. These further dimensions of individuality are subject to additional forms of constitution. I want now to suggest skeletally how the account presented in the previous two sections fits into a wider story of the social constitution of the individual. I will not sketch this wider story. Drawing on analyses of

Michel Foucault and Judith Butler, I will instead simply identify four further types of social constitution and relate them to three additional dimensions of individuality.

Two of these types of constitution can be described via Foucault's account of modern sexuality. The first is "incitement." Incitement occurs when a matrix of doings, sayings, and explicit propositional entities fastens upon, singles out, or calls attention to some already existing feature of human bodily existence and thereby intensifies, consolidates, transforms, or draws that feature out. In his analyses of sexuality, Foucault describes how incitement can take a variety of forms, from a simple strengthening of pleasures, as occurred with the pleasures of masturbation following increased attention to them, to the transformation of sexual pleasures into an overpowering presence and concern in life, as occurred with the sexualizing of the child's body. In these and other cases, particular words, ways of speaking, and practices such as confession and observation focus people's attention upon their bodies; and this increased attention, in the emotionally, ethically, spiritually, and scientifically charged atmosphere in which it occurs, induces a transformation in some physically based and socioculturally nurtured aspect of bodily existence. In constitution as incitement, consequently, "constitution" refers to the emergence of a new person through transformed functions, pleasures, sensations, and/or acts.

Although incitement can affect many dimensions of individuality, Foucault's analyses highlight its contribution to the constitution of two such dimensions: person type and subjecthood. By "person type" I mean the type of person one is, for example, type A individual, hyperactive child, nymphomaniac, homosexual, or delinquent. By "subjecthood" I mean, in Foucault's words, "be[ing] tied to one's identity by conscience or self-knowledge."[26] Incitement recasts the types of person one is by underwriting changed pleasures, attitudes, and behaviors. It transforms subjecthood, moreover, by working through attention and thereby altering conscience and self-knowledge.

The social formation of expressive bodies described in the first section is a type of production that closely resembles incitement. Both forms of constitution encompass the grasping and shaping of raw material. One difference is that incitement transforms already existing features, while production brings them into being. Another is that incitement involves attending to and transforming bodily existence, whereas production denotes placing a body in particular social settings and contexts, as a result of which either the person concerned becomes someone he or she was not before or there comes to be a person associated with that body in the first place.

A second further type of constitution underscored in Foucault's histories does not, like incitement and production, bring about new objects. It instead consists in (re)conceptualizations that, in (re)organizing people's understanding of reality, serve as axes around which they think about and interact with the world differently. The novelty such reconceptualization induces lies solely, at least initially, in changed ideas along with the accompanyingly reorganized thoughts and behaviors. Constitution as conception can effect significant transformations in subjecthood and person type. Consider Foucault's analysis of sex.

Sex, Foucault writes, is not a real entity. Rather, it is a discursively cobbled-together artificial unity, which in the nineteenth century (and still today) was conceived as a reality, although it really only represented a collocation of anatomical elements, biological functions, conducts, sensations, and pleasures.[27] (It is only in retrospect, of course, that revisable judgments can be pronounced about whether particular conceptual novelties such as sex are fictitious constructions or discoveries of important truths.) *Qua* entity, moreover, it was treated as a polymorphous causal principle that lies behind rational consciousness and is responsible for pleasures, desires, behaviors, and physiological events. It was even considered the truth or essence of the human soul. Individuals in the nineteenth century who accepted sex as an underlying causal and essential entity thought about and acted toward themselves and others differently, thereby transforming both their subjecthood and the person types they exemplified. Sexuality, for instance, became a prominent component of identity and knowledge of it a central moment in subjecthood; individuals, moreover, became hysterics or nymphomaniacs. Today sexuality remains a crucial component of subjecthood and person type, even as the identities, knowledges, and types involved change.

Two further additional kinds of constitution are prominently displayed in Judith Butler's analysis of a different component of individuality, gender. On Butler's "performance" view, gender identity is a bodily enactive combining of four types of element: sex, gender, sexual practices, and desire.[28] "Sex" refers to significant anatomy, to bodily parts with particular significances. (This anatomy, however, is not necessarily fixed or binary in sexual possibilities.) "Gender," for my purposes, can be understood as an array of identifications (adopted via fantasy) that orients how someone interacts with and carries herself in the world. "Sexual practices," meanwhile, refers to the sexual behaviors and acts a person performs, whereas "desire," finally, means something like a positive wanting to have or enjoy something. An individual with a specific gender comes to exist through the performative combination of these factors – thus, by

virtue of a body, with certain significant anatomical features, performing particular behaviors that express orientating identifications
as well as desires for particular objects and persons.

Two features of this performance theory are of importance in the
present context. First, gender identity is a "corporeal style" instituted
in a stylized repetition of acts.[29] It is a way of bodily being, of bodily
comportment. All four elements of gender, in other words, and not
anatomy and acts alone, pertain to the bodily. As in Wittgenstein,
desires and orientating identifications are exhibited in both what a
body says and how it carries out certain actions, gestures, and movements. Desires and identifications are not attributes of an inner substance or realm that give rise to "outer" behavior, but ways of being
signified by the stylized repetition of acts. Second, not any possible
combination of sex, gender, sexual practice, and desire is permitted.
Social regulatory practices prescribe certain combinations and punish
individuals who adopt illicit ones. The space of gender possibilities
exhibits a range of enjoined combinations constituting a variety of
straight men and women, a spectrum of sanctioned combinations
that constitute homosexuals of different types, and an *Abgrund* of
unintelligible combinations representing unclassifiable individuals.

The first of the two further types of constitution highlighted by
this theory is delimitation. Delimitation is the practical-discursive laying down of a realm of possibilities (in Butler's case, possible gender
positions). Such realms are neither fixed nor delimitable. As practices
evolve, so too do possibilities; and it is never possible even in principle
to draw up a complete inventory of a field's contents. The second
type of constitution is enforcement, the realization of particular possibilities through normative sanction and coercion. That most bodies
perform syntheses constituting heterosexual men and women results
from the enforcement of gender norms. The social "constitution" of
gender consists, as a result, in discursive practices opening a range of
stylized behavioral syntheses while normatively training and disciplin-
ing bodies to enact certain ones. I should add that at the core of
Butler's conception of gender identity lies the notion of institution
that figures prominently in my account: a body's performative syn-
thesis of anatomy, orientation, behavior, and desire *institutes*, that is,
establishes that there is an individual of such and such a gender.

The social constitution of the individual thus embraces at least six
types of constitution: production, institution, incitement, conception,
delimitation, and sanctioning. The Wittgensteinian account of social
constitution sketched in the current chapter most prominently applies the notions of production and institution to mind/action/body:
An expressive body is produced from a biological body by enclosure
within social practices; and a person's life conditions are instituted

by his or her contextualized behaviors and inner episodes on the background of the conceptual understandings carried in those practices. Sanctioning, too, however, appears in this account with the teaching and correction endemic to social practices. And, as we shall see in the following chapter, mind/action is also socially delimited, since practices establish the range of conditions open to people. My Wittgensteinian account of constitution, consequently, has not yet been fully explicated.

I wrote in the previous chapter that mind/action is the core component of individuality. All other dimensions of this status, I claimed, presuppose it. This holds for the additional components of individuality touched upon in this section. For example, both a person's gender and the person types she exemplifies rest squarely on her attitudes, behaviors, desires, and so forth. In any event, since mind/action forms the core of personhood, the social constitution of the individual consists most centrally in the production of an expressive body within sanctioning social practices, which practices, by establishing the possibilities of while also helping to institute on particular occasions that such and such doings, sayings, and inner episodes express specific life conditions, make it the case that there is someone in those conditions expressing them through this body.

Coda: What Is Mind? (II)

Drawing on a phrase of Klages, I wrote at the conclusion of the previous chapter that, for Wittgenstein, bodily activity is the appearance of mind and mind the expressed of bodily activity. The current chapter showed that bodily expression and mind are both social. The earlier formulation, therefore, can be elaborated to read that bodily activity is the socially molded appearance of mind and mind the socially instituted expressed of the socially molded body. Or more fully: Mind comprises the life conditions made present in the activities of socially molded expressive bodies and both instituted and made possible by participation in extant social practices.

4

Social Practices

The previous chapter argued that mind/action is socially instituted and the body expressing it socially produced. It is within the context of the reactions, attributions, teachings, and corrections of others that a person comes coordinately to express and be in life conditions, to ascribe them to herself, and to grasp them in others. I glossed this position by claiming that these matters come to pass through participation in social practices. Yet I said nothing about practices. The account of social constitution, consequently, remains incomplete. The current chapter rectifies this situation, while also providing basic elements of the account of sociality developed in Chapter 6.

Wittgenstein wrote almost nothing directly concerning the nature of social practices. I still want, however, to claim the appellation "Wittgensteinian" for my analysis. It approaches the topic of practice using the preceding Wittgensteinian account of mind/action and its social constitution as a platform and occasionally touches on specific statements and views of his. At the same time, it largely leaves Wittgenstein's remarks behind. Thus in contrast to the previous analysis of mind/action, which claims the label "Wittgensteinian" *qua* systemizing interpretation of his late writings, the impending account of practice claims this label only *qua* analysis partly based on and inspired by his remarks.

By tackling an issue, the nature of social practices, that is not a prime topic of the remarks from which it takes its orientation, my account enters terrain already explored by the stream of contemporary thought called "practice theory." My exposition in the current chapter, while turning away from Wittgenstein, turns toward and appropriates analyses of practice complementary to a Wittgensteinian approach, in particular several cultivated in or shaped by Heidegger's *Being and Time*. I will hold off until Chapter 5 a confrontation with the most extensively worked out accounts of practice, those of Bourdieu, Giddens, and Lyotard. It is possible fully to appreciate and engage their ideas only from the perspective of a well-developed alternative.

Three Notions of Practice

At least three notions of practice are prominent in the current conjuncture. According to one, practicing is learning how or improving one's ability to do something by repeatedly working at and carrying it out. It is in this sense of practice that adults practice the piano, children are enjoined to go practice, and beginners as well as adepts inform their friends on the phone that they can't because they're practicing. This notion of practice is not irrelevant to the current discussion. But however vital and practically necessary practice *qua* development through doing may be, it will be put aside in what follows. I want instead to concentrate on two other notions.

The first is practice as a temporally unfolding and spatially dispersed nexus of doings and sayings. Examples are cooking practices, voting practices, industrial practices, recreational practices, and correctional practices. To say that the doings and sayings forming a practice constitute a nexus is to say that they are linked in certain ways. Three major avenues of linkage are involved: (1) through understandings, for example, of what to say and do; (2) through explicit rules, principles, precepts, and instructions; and (3) through what I will call "teleoaffective" structures embracing ends, projects, tasks, purposes, beliefs, emotions, and moods. All three will be extensively discussed in the following.

These avenues are not, of course, the only forms of linkage among actions. Doings and sayings also, most importantly, form causal chains, each link of which consists in the succeeding action responding to the previous one (or to a change in the world the previous action brought about).[1] Many of the sayings and doings that make up a practice are linked as cause and effect. Causal connections are part of a practice, however, only if they somehow "result" or "follow" from the practice's organization, that is, the understandings, rules, and teleoaffective structure linking the practice's constituent actions – and many of the causal connections among a practice's actions do not follow from its organization. What's more, actions that appear in different practices can also be causally connected. In fact, actions collect through causality into various sorts of spatiotemporal network that, in running through and connecting different practices, must be captured by terms other than "practice" – for example, "institution," "group," and "system" (see Chapter 6).

Perhaps it will initially help clarify matters to point out that practices, in Jürgen Habermas's terminology, are something through which "social" as opposed to "systems" integration is primarily achieved. Social integration is the coordination of actions through the harmonization of "action orientations," whereas systems integration is the coordination of actions through the functional intermeshing (*Vernetzung*) of the unintended consequences of action.[2] The organization of a practice principally coordinates actions by organizing factors that govern participants' actions, and not by stabilizing the unintentional consequences of these actions (although this organization, as just indicated, also establishes chains of action). I add that even though the organizational significance of practices lines up with his categories thusly, Habermas's specific analysis of social integration does not adequately capture the "harmonization" of lives within practices. For instance, in claiming that action orientations are harmonized through normatively guaranteed or communicatively achieved consensuses, he overlooks the prior and enveloping "harmonization" effected through the articulation of intelligibility. Moreover, in conceiving social integration as the harmonization of "lifeworlds" and lifeworlds as essentially knowledge (practical and propositional), Habermas's analysis does not easily accommodate the fact that social integration is secured within phenomena (practices) that are spatiotemporal entities and not merely configurations of such action-determining components of knowledge as goals, themes, plans, and definitions of the situation.[3]

A third prominent notion of practice is that of performing an action or carrying out a practice of the second sort. This notion denotes the do-ing, the actual activity or energization, at the heart of action. It is this notion of practice that the Western philosophical tradition has often opposed to theory: theory versus practice, contemplation and reflection versus doing. Of course, since contemplation and reflection are themselves activities that contemplators and reflectors carry out, the opposition, as sometimes remarked, is better described as the segregation of one form of activity from all others. In any event, this third notion of practice is central to any analysis of human existence. It designates the continuous happening at the core of human life *qua* stream of activity and reminds us that existence *is* a happening taking the form of ceaseless performing and carrying out. This notion of practice also closely connects with the second. Each of the linked doings and saying constituting a practice is only in being performed. Practice in the sense of do-ing, as a result, actualizes and sustains practices in the sense of nexuses of doings. For this reason, a general analysis of practices *qua* spatiotemporal entities must embrace an account of practice *qua* do-ing; in more standard language, it must offer an account of action. This thesis is confirmed by the theories appearing in this and the subsequent chapter.

Dispersed Practices

I label a first category of spatiotemporal practice "dispersed" to emphasize that practices of this sort, in contrast to those of a second category I will call "integrative," are widely dispersed among different sectors of social life. Examples of dispersed practices are the practices of describing, ordering, following rules, explaining, questioning, reporting, examining, and imagining. As Wittgenstein writes, "To obey a rule, to make a report, to give an order, to play a game of chess, are *customs* (uses, institutions) [*Gepflogenheiten, Gebräuche, Institutionen*]" (*PI*, 199). The dispersed practice of X-ing is a set of doings and sayings linked primarily, usually exclusively, by the understanding of X-ing. This understanding, in turn, normally has three components: (1) the ability to carry out acts of X-ing (e.g., describing, ordering, questioning), (2) the ability to identify and attribute X-ings, in both one's own and other's cases, and (3) the ability to prompt or respond to X-ings. Another expression for "the ability to" in this context is "knowing how to." Since a practice's doings and sayings express the understanding composing the practice, a dispersed practice is a set of doings and sayings linked primarily by an understanding they express.

Unlike the first two components of this understanding, the third is not an omnipresent feature of dispersed practices. It is present whenever established ways of prompting or responding to a given sort of action exist. Examples are the practices of questioning, ordering, greeting, describing, and reporting, where answering, obeying, returning greetings, and asking for a description or report, are established promptings or responses. No one who understands questioning, ordering, and so on fails to grasp this. A practice where the third component of understanding is absent is following a rule. Notice, furthermore, that established promptings of and responses to X-ings, as well as acts of identifying and attributing X-ings, often help compose further dispersed practices (e.g., inquiring, answering, and obeying). This indicates that dispersed practices, including their constitutive understandings, are woven into nexuses. They are not isolated and self-contained atoms. Their "dispersion" consists simply in their widespread occurrence across different sectors of social life, a breadth that, as noted, helps distinguish them from integrative practices.

Dispersed practices are only governed very rarely by the second and third forms of linkage among practice-composing behaviors: rules and teleoaffective structures. Their constituent actions are held together mostly by understandings and need not either lean on principles, instructions, and pointers or be informed by particular orders of ends, purposes, beliefs, and emotions. This fact, too, helps distinguish dispersed from integrative practices. The dispersion of a practice actually requires the absence of the teleoaffective component of practice organization. If the doings and sayings that make up such practices as describing, explaining, and questioning were tied to particular arrays of ends, purposes, emotions, and the like, the practices could not appear in more or less all walks of life and in a wide variety of situations.

Now, an important feature of action – one that will figure prominently in Chapter 6 – is that individual acts of X-ing presuppose the dispersed practice of X-ing. The argument for this claim runs as follows. As explained in the previous chapter, a doing or saying constitutes an X-ing because of the contexts in which it is performed and against the background of the understanding of X-ing (and Y-ing and Z-ing, etc.). I have now explained that the dispersed practice of X-ing is a set of doings and sayings linked (primarily) by the understanding of X-ing. It follows that a doing or saying constitutes an X-ing on the background of an understanding carried by the practice of X-ing. Any given X-ing, consequently, presupposes the practice.[4]

It might appear that this brisk and abstract argument rests on a sleight of hand. For the understanding of X-ing at work in the previous chapter, the understanding of the concept of X-ing, is manifested both in uses of the word "x" and in reactions to the phenomena thereby qualified. It embraces, therefore, the abilities to identify/attribute X-ings and prompt/respond to them. These abilities, however, are only two of the three constitutive of that understanding of X-ing which was claimed in the current chapter to link the doings and sayings composing the practice of X-ing. The argument, consequently, might seem to rest on an equivocation.

The analytic disparity between the two-part conceptual understanding of X-ing spoken of in the previous chapter and the three-part understanding introduced in the current one is real. However, the conceptual understanding of, say, describing, which is expressed in attributions of and responses to acts of describing, still presupposes the practice of describing. For this conceptual understanding is established, acquired, employed, and thus sustained amid acts of describing and attributions of as well as responses to such acts.[5] The conceptual understanding of describing could be disengaged from the practice of describing only if it could be formulated (by participants) in sentences furnished to nonparticipants, who would then enjoy a nonparticipatory propositional understanding of describing. This conceptual understanding is not, however, fully formulable in words. Propositional understanding is an inferior stand-in for, and thus not equivalent to, the nonpropositional conceptual understanding living in the practice. (This claim in no way impugns the capacity of all sorts of formulation to effect changes in nonpropositional understanding.) So the type of understanding discussed in the previous chapter cannot be disengaged from the practices that sustain it. And even if it could, since the propositional understanding is an articulation *of* the living one, the propositional understanding would presuppose it and the practice in which it is embedded. So the conceptual understanding of X-ing spoken of in the previous chapter presupposes the practice of X-ing. Any instance of X-ing, consequently, also does so.

More or less the same argument applies to those practices in which, due to the cultivation of highly specialized skills, some participants acquire a conceptual understanding expressed in recognitions and responses that is thinner than the understanding expressed in recognitions, responses, and performances attained by others. Someone, for instance, can recognize flute playing and even respond appropriately to it without being able to play the flute. Again, however, this understanding of flute playing exists as a matter of fact only in the context of the practice of flute playing where it is established, employed, and sustained. The nonplayer acquires this understanding, moreover, through live as well as linguistically and electronically mediated participation in and exposure to the practice, wherein performances are encountered. Once again, nonparticipants (and neophytes) can be furnished verbal specifications in lieu of and/or prior to participation. But although such propositional understanding might enable them to call the same behavior flute playing as others do, it will probably not enable them to join in with responses to such playing. And, in any case, since the propositional understanding articulates, it thus presupposes the understanding that exists within the practice. The conceptual understanding of flute playing thus presupposes the practice thereof. Any instance of flute playing, consequently, does so too.

This conclusion generalizes to all actions: The (conceptual) understanding, against which a particular behavior-in-circumstances constitutes X-ing, is carried by the practice of X-ing.[6] This should not be a surprising result. It is only to be expected that identity-bestowing understandings of action inhabit and thrive within the manifolds of doings and sayings in which the identities involved take hold. Notice, moreover, that this argument helps substantiate a point gestured at in a footnote in the previous chapter, namely, that the conceptual understandings against which behavior-in-circumstances expresses specific life conditions are carried in practices.

This interweaving of understanding and activity underlies an existential dilemma that many anthropologists report facing during fieldwork: to go or not to go native. What people are doing when they behave in an unfamiliar and not readily intelligible way often depends on an understanding that is expressed in their behavior and that links this behavior with others. Manuscripts, films, lectures, and discussions can convey some of this understanding to an anthropologist prior to her arrival in native lands. How well they accomplish this depends in part on the sensitivity, reflectivity, care, and agility with which the other anthropologists crafting these artifacts and conducting these activities articulate that grasp of this understanding which they have acquired through exposure to and participation in native practices. If the fledgling anthropologist is to sharpen and extend their grasp, she too must be exposed to and participate in these practices; she must to some extent become "one of them." Since the more unreservedly she does this the profounder her grasp of the sought understanding becomes, one of her professional raisons d'être counsels her to resolutely go native. Doing this, however, harbors cultural and psychological dangers, for instance, a turbulent identity in disarray. The resulting dilemma thus reflects the intimate entanglement of understanding and doing.

In conclusion, the conceptual understanding of X-ing, on the basis of which behavior-in-circumstances constitutes X-ing, is not other than but *part* of that understanding of X-ing which links and is expressed by the behaviors forming the practice of X-ing. X-ings, consequently, presuppose the practice of X-ing. As Wittgenstein writes,

Is what we call "obeying a rule" something that it would be possible for only *one* man to do, and to do only *once* in his life? . . . It is not possible that there should have been only one occasion on which someone obeyed a rule. It is not possible that there should have been only one occasion on which a report was made, an order given or understood; and so on. To obey a rule, to make a report, to give an order, to play a game of chess, are *customs* (uses, institutions). (*PI*, 199)

Incidentally, Malcolm Budd has pointed out that in this passage Wittgenstein does not explicitly claim that a condition of the actions mentioned is the existence of a social practice, only that any one of them presupposes other actions of the same type. All the actions of a given type might, therefore, be performed by the same person.[7] The passage thus leaves open whether the "practice" presupposed by an action is an individual or social one. Denying, however, that Wittgenstein thought of the presupposed practice as an actual social one is highly problematic. In the first place, the terms *Gepflogenheiten*, *Gebräuche*, and *Institutionen* (though not necessarily *Praxis* or the first

two terms in the singular) connote sociality.[8] Equally important, as
that, the mastery of which would qualify different behaviors of a
single person as actions of a particular type and thereby bind them
together into an individual practice, Budd suggests "techniques" and
"methods of use." Surely, however, Wittgenstein thought of these as
social in the sense of mastered essentially by multiple people.[9] As
Budd indicates, Wittgenstein's teaching, that meaning lies in use and
not in something that at the time of speaking or writing either takes
place or comes before the mind, does not in itself imply that use is
individual or social. Given, however, the considerations pertaining to
recognition and identity raised in the private language argument,[10]
the "community" interpretation is hardly disapproved by merely
pointing this fact out – even in conjunction with the justified dismissal of dubious interpretations such as Christopher Peacocke's
(which construes the communities whose practices are presupposed
by individuals' actions as possible and not actual communities).[11]

   Since I am appropriating the designation Wittgensteinian for my
account of practices, I should briefly relate the notion of a dispersed
practice to that of a language-game. Like "form of life," "language-
game" does not so much pick out any specific type of entity as
crystallize a general viewpoint toward language. In Wittgenstein's
(late) writings, consequently, the term is applied to a remarkably
diverse range of human goings-on with language, including writing
and not only speaking (*RPP* I, 933); "primitive" activities with particular words (*PI*, 7); activities involving the use of several words (*PI*, 7;
Z, 644); games with just one player (*OC*, 255); and affairs in which
language plays a minor role (*RPP* I, 945). Invented as well as fantastic
activities, affairs, and proceedings can also be language-games. The
attitude, on the other hand, that the term expresses is clear-cut:
"Hence the term 'language-*game*' is meant to bring into prominence
the fact that the *speaking* of a language is part of an activity" (*PI*, 23).
A language-game is an open-ended set of behaviors into which the
speaking and using of language are woven (cf. *PI*, 7, *OC* 519).

   Dispersed practices of X-ing are language-games in this vague
sense, though saying this adds nothing to our appreciation of them.
Conversely, many of the language-games Wittgenstein discusses qualify as dispersed practices. An example is a type of language-game
that Wittgenstein is particularly fond of citing, namely, the language-
game with the word "x." The language-game with the word "x" is
composed of uses of the word together with actions (and phenomena) that either occasion those uses or respond to those uses or to
the phenomena they qualify.[12] It comprises, consequently, a set of
behaviors linked by the understanding of X, and thus can be added
to the roster of dispersed practices. Wittgenstein also sometimes re-

fers to what I initially identified as dispersed practices (e.g., the practice of reporting) as language-games (e.g., *PI*, 23; *LW* II, p. 8c). However, since Wittgenstein calls a large variety of goings-on language-games, some of austere simplicity, not all language-games as he uses the term are dispersed practices.

I want now to draw out additional features of dispersed practices by explicating how they fall beneath the threshold of what Michael Oakeshott calls practices in his ingenious and provocative analysis of the topic. Oakeshott defines a practice as follows:

A practice may be identified as a set of considerations, manners, uses, observances, customs, standards, canon's maxims, principles, rules, and offices specifying useful procedures or denoting obligations or duties which relate to human actions and utterances. It is a prudential or a moral adverbial qualification of choices and performances, more or less complicated, in which conduct is understood in terms of a procedure.[13]

A practice is a set of considerations that governs how people act. It rules action not by specifying particular actions to perform, but by offering matters to be taken account of when acting and choosing. When observed, consequently, it qualifies the how as opposed to the what of actions. For instance, the words "civilly," "punctually," "scientifically," "legally," "morally," and "poetically" do not specify particular substantial actions. The practices of civility, science, law, morality, and poetry for which they stand are sets of considerations and procedures, which if observed qualify whatever is done as civil, punctual, scientific, legal, moral, or poetic (p. 56). The practice of morality, for example, is a "vernacular language of colloquial intercourse" (p. 63), which offers actors notions, standards, uses, rules, and maxims of good and bad conduct that they may or may not observe in acting and deciding what to do. Practices, moreover, can range from relative plasticity to "the firmness of an 'institution' "; they can overlap, form hierarchies, and join to compose more complex practices; and a given performance or choice can be governed by a variety of them (p. 56).

What a person does on any occasion depends not on practices, but instead on (1) the understanding he has of his situation, which in diagnosing this situation as somehow unacceptable, specifies a wished-for condition that remedies the unacceptability and that the action performed is designed to bring about (pp. 38–39); and (2) the motives ("sentiment") out of which a particular action conducive to the wished-for condition is "chosen."[14] An action, consequently, is a motivated response to an unacceptable situation that aims to bring about a condition relieving this unacceptability. Practices simply provide considerations the actor can observe or ignore in reaching a choice of alleviating action.

As Oakeshott indicates, practices *qua* considerations house a form of social association: Individuals associate by virtue of subscribing to one and the same practice. I applaud Oakeshott's asseveration that practices form the baseline of human sociality (see p. 87). I want, however, to rebuff his attempt to carve off practices from the actions they govern. Oakeshott sunders them because, as a good conservative, he wants the identity of action to derive solely from features of individuals. This he achieves by tethering what someone does to his or her understandings and motives. Once, however, the identity of action is pegged to the individual, practices (i.e., sociality) can only pertain to the how of action. "[N]o action or utterance is what it is in terms of its subscription to a practice" (p. 98). *How* it is might be so bound; and the how and the what of an action might be "inseparably joined" (p. 57). But action is essentially an individualist phenomenon only contingently related to sociality. In my account, on the other hand, an action is the action it is as part of a practice. Action, consequently, is essentially social; and practices must consist not only in considerations but also in actions.

The unpropitiousness of Oakeshott's fission is revealed by problems that assail the relation he envisions between actions and practices. He writes that the considerations that form a practice are not "applied" in acting since they do not specify substantial actions (p. 90). They are instead "used" (p. 120). Of many types of consideration that compose practices, however, it is untrue that people use them. Standards, rules, principles, and above all manners (a true adverbial qualifier) might be "used," but this cannot be said of uses, observances, and customs. These are "carried on," "upheld," or "continued." They "qualify" a performance when it carries them out. Oakeshott has thus assembled a motley crew of "considerations" with dichotomously disparate bearings on action.

Further problems beset the thesis that the considerations that constitute practices cannot specify substantial actions. The best illustration Oakeshott provides of this infirmity involves such moral precepts as "treat neighbors kindly." On his analysis, this precept does not prescribe particular actions, but instead enjoins an actor to consider what treating neighbors kindly amounts to in particular situations. Of course, if observed, the precept *does* specify at a certain level of description what someone does, namely, "treats neighbors kindly." This raises the question of at what level of description is what someone does a "substantial" action, an issue Oakeshott fails to confront. In addition, not all rules and principles that constitute practices share the lack of specificity exhibited by this moral precept. For instance, the banking rule, Lock the vault at 4 P.M., specifies locking the vault at 4 P.M. To perform this action, an employee must of course perform other actions that in the situation constitute locking it. Unlike vis-à-vis treating neighbors kindly, however, neither the banking rule nor the employee's understanding and motives (in Oakeshott's sense) need specify a description of those other actions. "Lock the vault" is already a description of sufficient specificity that it *can* be carried out by the performance of bodily doings and sayings. As a result, the employee, without the intercession of understanding and motives, might simply perform behaviors in her bodily repertoire that constitute locking up in that situation (e.g., calling out "clear the way" and pushing a button). If so, then the banking rule, and not understanding and motives, specifies what she does. The what of action cannot, therefore, be categorically shielded from practices, even on Oakeshott's construal of the latter. Oakeshott could hope to drive a wedge between practices and actions only by artificially restricting the rules, precepts, and principles that constitute practices to ones of relative generality.

Such nonexplicit "considerations" as uses and customs also, however, specify what people do. For instance, the custom of extending one's hand when meeting an acquaintance on certain occasions specifies what certain people do as exactly as motivation plus understanding could. The "observance" of the custom does not consist in taking particular considerations into account, but instead in performing certain specific actions and thereby carrying the custom forward. Oakeshott's attempt to split practice from action is thus rendered otiose by the fact that the nonpropositional elements of his practices (customs, uses, observances) generally determine the what of action. Note that Oakeshott's talk of "using considerations" fails to do justice to the nonpropositional dimension of practice. Although his customs, uses, and observances are coextensive with what I label the understandings that imbue dispersed practices, nonpropositional understanding appears in his account solely as the adept's ability to employ the considerations of practices in performing and choosing actions (cf. p. 91).

Integrative Practices

By "integrative practices" I mean the more complex practices found in and constitutive of particular domains of social life. Examples are farming practices, business practices, voting practices, teaching practices, celebration practices, cooking practices, recreational practices, industrial practices, religious practices, and banking practices. Like dispersed practices, integrative ones are collections of linked doings and sayings. The doings and sayings involved are joined by: (1) understandings of Q-ing and R-ing (etc.), along with "sensitized" understandings of X-ing and Y-ing (etc.), the latter carried by the transfigured forms that the dispersed practices of X-ing and Y-ing adopt within integrative practices; (2) explicit rules, principles, precepts, and instructions; and (3) teleoaffective structures comprising hierarchies of ends, tasks, projects, beliefs, emotions, moods, and the like. I will henceforth refer to the understandings, rules, and teleoaffective structure that link an integrative practice's doings and sayings as the "organization" of the practice.

It is important not to think of integrative practices as assemblages of dispersed practices, which are added together to form integrative ones (in which case dispersed and integrative practices would stand in the relation of element to complex). Although multiple dispersed practices wander through and meet within integrative ones, they are sometimes transformed through their incorporation. The activity of questioning, for instance, comprises different doings and sayings and is imbued with different understandings within legal, religious, and interrogation practices. As a result, the asking and answering of questions in these practices are not the carrying on of a dispersed practice. In addition, not all simpler activities that help compose integrative practices are former dispersed practices incorporated into integrative ones. Integrative practices often establish activities (e.g., marking ballots) that enjoy the relative simplicity of dispersed practices and are linked primarily by understanding alone, but exist only within the compass of the integrative practices concerned (although they might break free to become dispersed practices, e.g., praying).

Conversely, one must not think that integrative practices alone exist and that, instead of such dispersed practices as describing and requesting, social life exhibits only the varying forms of describing and requesting found in different integrative practices. Some relatively simple activities retain the same form in different integrative practices and are not, therefore, nothing more than the varying forms they take within these practices. For instance, such dispersed practices as describing and requesting course largely unaltered through numerous domains of social life, regardless of how much they might be transformed within specific domains. What is true is that people usually, though not always, are also engaged in an integrative practice when carrying on a dispersed one. When someone describes something, for example, he or she is usually also carrying on farming, nautical, cooking, education, military, or building practices. An act of describing, consequently, can be part of both the dispersed practice of describing and a particular integrative practice. This does not, however, negate the existence of the dispersed practice – that is, the distinctness of an array of doings and sayings linked by the understanding of describing, together with the independence of this linked manifold from particular integrative practices. The nexus of practices that is the social field embraces dispersed as well as integrative practices.

The "understandings of Q-ing and R-ing" referred to in the preceding specification of the organization of an integrative practice are the understandings that link the doings and sayings composing the relatively simple activities (Q-ing and R-ing) established within such practices. They are not, therefore, understandings that compose dispersed practices.[15] By "sensitized" understandings (of X-ing and Y-ing etc.), meanwhile, I mean the abilities to act and speak in the ways characteristic of the transformed forms assumed by such dispersed practices as ordering in particular domains of life. When someone enters, say, military life, the understanding of ordering he has previously acquired becomes sensitized to the particular way the activity runs on there: the particular doings and sayings that constitute issuing and acknowledging orders; the appropriate times and places to issue and respond to them; and the rhythm and temporal pace[16] as well as tones, particular words, and gestures that qualify acts of ordering as dominating, disdainful, or hesitant and certify acts of acknowledging orders as insolent, disrespectful, or obeisant.

The doings and sayings of an integrative practice are also linked by explicit rules, principles, precepts, instructions, and the like. This means that people take account of and adhere to these formulations when participating in the practice. Oakeshott's account of "using" explicit considerations here finds its home. As noted, however, rules, principles, maxims, and the like can, *pace* Oakeshott, specify particular actions. Think, for instance, of cooking instructions. Explicit formulations "govern" action not only when someone performs whatever it is that he thinks observing them amounts to in a specific situation, but also when he or she performs the specific actions they enjoin because they are enjoined.

In writing that "teleoaffective structure" further links the doings and saying of an integrative practice, I mean that these behaviors express hierarchized orders of ends, purposes, projects, actions, beliefs, and emotions that fall within a certain field of possible such orders. (Keep in mind that, throughout this and the following chapters, "actions" includes speech acts.) The delimitation of this field will be discussed momentarily. Unlike explicit rules, the orders constituting a teleoaffective structure need not be spelled out and explicitly enjoined in formulations, although formulation does sometimes occur, especially (but not only) in learning situations, in the face of nonstandard doings and sayings, and on the occasions when the flow of reactions suffers what Hubert Dreyfus calls "breakdowns" in continuous absorbed coping.[17] Western cooking practices, for instance, embrace a range of ends, projects, purposes, and tasks such as making snacks, lightly browning, whipping up appetizers, chopping, pickling, readying the grill, and preparing healthy meals. These ends and projects form myriad possible hierarchical orders in the sense that certain actions and projects can be carried out for the sake of certain other projects and ends. Even though a person cannot cook without pursuing some subset of these orders, the ends, projects, and hierarchies involved may or may not be prescribed and/or even described by cooks, cookbooks, and parents. Notice that the greater part, sometimes the entirety of a practice's teleoaffective structure concerns teleology alone instead of also or exclusively affectivity. No particular emotions and moods, for instance, are appropriate for cooking. By contrast, certain emotions are appropriate for, and nearly inherent to rearing practices, for example, love, hate, and affection in Western versions. Think also of celebration and religious practices and of that miscellany of practices that anthropologists label "rituals."

A teleoaffective structure is in fact a collection of possible orders of life conditions. This is obvious vis-à-vis its constituent beliefs, actions, emotions, and moods. It also holds of its ends, purposes, projects, and tasks. What it is for a person to pursue ends and purposes is for the sought states of affairs to be objects of her intentions, desires, hopes, and wants. (I will return to this point.) Projects and tasks, meanwhile (e.g., making dinner and preparing healthy meals), are simply actions of sufficient generality that they are carried out or effected through the performance of other actions, themselves performed via bodily doings and sayings.

Now, the possible hierarchies of ends, projects, and so on composing a practice's teleoaffective structure are those that are normative for participants in the practice.[18] As is immediately obvious, however, the entirety of a practice's organization is normative. By "normativity" I mean, in the first place, oughtness or rightness. The understandings, rules, and teleoaffective structure that organize a practice specify how actions (including speech acts) ought to be carried out, understood, prompted, and responded to; what specifically and unequivocally should be done or said (when, where . . .); and which ends should be pursued, which projects, tasks, and actions carried out for that end, and which emotions possessed – when, that is, one is engaged in the practice. Evidence for a practice's organization is thus found in the presence and absence of corrective, remonstrative, and punishing behaviors and in the verbal and nonverbal injunctions, encouragements, and instructions whereby neophytes are brought into line. (Further evidence is found in the prevalence of particular actions and combinations of conditions.)

By "normativity" I mean, second, acceptability. A practice's organization establishes not only that certain actions are correct (in certain situations), but also that other actions are acceptable, even if they are not how one should proceed. The understanding of X-ing, for example, opens a range of ways of acceptably X-ing as well as prompting and responding thereto. A person who understands ordering in military practices grasps not only correct procedures for acknowledging the orders of superiors, but also a range of actions that can be performed in response to the issuing of orders without incurring correction, remonstration, and punishment. A teleoaffective structure similarly embraces a field of acceptable orders of life conditions that is wider than whatever range of orders is marked as correct. Eating practices, for instance, might embrace several acceptable sequences for consuming the courses of a meal instead of laying down one right order. Business practices might likewise refrain from promoting one end as the single right one (e.g., maximizing profits) and instead acknowledge several acceptable ones. And farming practices might not single out one correct network of projects and tasks to perform during the growing season, but instead open up a number of acceptable configurations thereof. In sum, the range of life condition orders embraced by a practice's teleoaffective structure is the range of orders that it is correct or acceptable for participants' behavior to express when participating in the practice.

I note that among the acceptable actions (and life condition orders) constitutive of a practice's teleoaffective structure are some that have not yet been carried out (or instantiated). Practices found possible novelty in that people happen upon new ways of proceeding, and others deem these ways acceptable, on the background of their participation in practices and familiarity with teleoaffective structures. The understandings that organize an integrative practice likewise, though more weakly, open ranges of acceptable doings and sayings broader than the behaviors already performed in the practice.

I should, furthermore, explicitly point out that, in claiming that understandings, rules, and teleoaffective structures organize integrative practices, I am in effect proposing that the notion of rules, with which some theorists have maintained that social activities generally are "rule governed," be split into three categories. In particular, insofar as rules have been widely construed as delimiting normativity, I am claiming that the establishment and constitution of rightness and acceptability in human practices is more perspicuously analyzed via this triad than via some generic type of rule alone. This means, inter alia, that allegedly rule-governed "customs" that are not governed by explicit rules (1) proceed on the basis of not fully formulable understandings of how to react and respond (e.g., the "custom" of shaking acquaintances' hands), and/or (2) conform to normative orders of ends and activities (e.g., the "custom" of correcting people's mistakes publicly for the sake of restoring order in public life). It also means, inter alia, that more complex activities governed by explicit rules also are governed by unformulable understandings and fields of correct/acceptable ends and activities. Insistence on the omnipresence of rules threatens to obscure not only the unformulability of understandings, but also the presence and complexity of normative teleological hierarchies. I should add that the proposed tripartite organization also contests those more recent theories that construe human activity as organized primarily or exclusively not by rules, but by understandings (know-hows). As will be discussed in the succeeding chapter, the inchoate nature of practical understanding raises problems for these theories.

An integrative practice is a set of doings and sayings linked by understandings, explicit rules, and teleoaffective structure. To say that these items link doings and sayings is not to say that the same understandings, rules, and structure govern all behaviors involved. Some behaviors are linked in being governed by the same components of practice organization, others via interrelations and cross-references among these components. A practice's behaviors are also often causally connected through its organization – for example, when it is by virtue of a person's understanding of describing that she responds to a request to describe something by describing it; or when it is correct within industrial practices in the car industry for a foreman to restart an assembly line both in response to an "All OK" signal and in order to resume (carrying out the project of) assembling chassis. Causal connections among a practice's constituent actions sometimes become regularized or prescribed – calling on students who raise their hands in class, handing over a slab when requested (*PI*, 2), taking evasive maneuvers in the face of an enemy patrol, and countering a right jab with a left undercut. Responses to such events as illuminating emergency lights, gathering storm clouds, ringing telephones, blowing whistles, and changing traffic signals also become regularized and normativized. As a result of not merely these regularized and normativized connections, practices exhibit elaborate causal chains of action. Causality also links actions of one practice with those constituting others, thereby resulting in enormous networks of action chains that run through and couple together integrative practices in convoluted and complicated ways (see Chapter 6).

Despite the complexity of these matters, as well as variation in the proportional mixes of the three components of practice organization that govern different integrative practices, the limits of any practice are drawn by its organization: A doing or saying belongs to a given practice if it expresses components of that practice's organization. Of course, insofar as a person's life conditions are indeterminate (cf. Chapter 3), which practice(s) a given doing or saying helps compose is also such. Further complicating the drawing of the boundaries of practices is overlap among integrative practices. A given action, for instance, might be part of two or more practices. When, for example, celebration practices are commercialized, some actions constitutive of celebrations also constitute business activities. Or a given rule or teleoaffective substructure might characterize one or more practices, for instance, the precept "respect your elders" or the procedure of gathering people together in order to make decisions. Additional complexities enter with the variation of practices across social space and history. Not only can dispersed practices, as discussed, be trans-formed within integrative ones, but such integrative practices as those of business, cooking, and rearing can vary across locales and among social formations (e.g., firms and families). The particular forms adopted by integrative and dispersed practices, including the under-standing, rules and teleoaffectivity organizing them, also evolve with time, often in response to events occurring within them. Indeed, such plural expressions as "teaching practices" and "farming practices" connote not only the varied activities that make up teaching and farming, but also the different forms teaching and farming take across social space and time.

People, it is important to note, are almost always – though not necessarily – aware of and also have words for the integrative prac-tices in which they participate. They are cognizant of such practices in part because with time the teleological structures and rules or-ganizing them come to their attention. An unrecognized integrative practice is much less likely than an unnoted life pattern (*Lebensmuster;* see Chapter 2) because a patterned string of spatiotemporally dis-persed behaviors-in-circumstances can be unobtrusive and insignifi-cant in a way that a nexus of in part causally connected actions organized by a cross-referencing and interlocked bundle of under-standings, rules, and teleoaffective structure cannot. Like words for life conditions, moreover, words for practices (e.g., "farming," "busi-ness," "celebration," "recreation," and "religion") are used in discus-sions and analyses of the world, thereby making younger people aware of commonplace practices, adults informed of obscure ones, and in this and other ways helping to perpetuate them. This means that a significant clue to which practices constitute people's lives is the vocabulary they use to classify their activities.

A final and extremely significant feature of integrative practices is that they are inherently social entities, that is, phenomena of human coexistence. As I will explain much more fully in Chapter 6, through participating in a practice a person *eo ipso* coexists with others, not merely those individuals with whom she interacts, but also wider sets up to and including the collection of all those party to the practice. For instance, she coexists with the set of participants whose behavior is governed by a particular component of the practice's organization that governs hers. She further coexists with indeterminate many others through the implicit or explicit references to people contained in some components, for example, the understanding of ordering, the rule "respect your elders," the project of educating children, and the end of upholding public image. These same considerations apply mutatis mutandis to dispersed practices. All integrative and dispersed practices are social, above all because participating in them entails immersion in an extensive tissue of coexistence with indefinitely many other people.

An integrative practice is social in the additional sense that its organization is expressed in the nexus of doings and sayings that compose it, as opposed to the individual doings and sayings involved. This is clearest vis-à-vis teleology and affectivity. Love among family members is expressed in particular doings and saying. But these behaviors express affection in part because they are performed in the context of one another: The manifold of acts of affection forms a context in which each individually expresses love. That which expresses familial love, consequently, is in a sense the manifold and not the acts taken individually. Similarly, a firm's employees perform actions that individually express particular components of the hier-archized field of ends, tasks, and purposes that characterizes business practices in the firm. Not only, however, does any one action do this only in conjunction with other actions' doing so, but the hierarchy is expressed *as* a hierarchy only in these actions taken as a set. This situation is reflected in the fact that the teleological organization is attributed to the firm, not its employees. Parallel remarks apply to the rules and understandings that organize integrative practices, since participants observe different subsets of a practice's rules and act out of different subarrays of its understandings.

It follows that the organization of an integrative practice is, in Charles Taylor's phrase, "out there in the practices themselves," as opposed to in here "in the minds of the actors."[19] As will be discussed in the upcoming section on intelligibility and language, Taylor con-strues what is out there in a practice as a field of "intersubjective meanings" that delimits how people and situations can make sense to participants. My claim is that "out there" in an integrative practice is the array of understandings, rules, and teleoaffective structure that organize it. This should not be thought of as "in the minds" of individual participants.

A parallel, but slightly weaker point holds of the understanding that organizes a dispersed practice. A person acquires this understanding through exposure to and participation in the practice whose actions express it. Once acquired, moreover, she perpetuates the practice by performing actions that signify the same understanding. The understanding, consequently, was and continues to be "out there" in the expanding manifold of behaviors. It is also, of course, "in her" in a way that the wider organization of an integrative practice cannot be. But it lodges there through her introduction into and exposure to past components of the continuing practice whose present constituent behaviors continue to express it. It is only because it is "out there" in something to which she becomes party that it is also "in her."

I also second Taylor's accompanying claim, that practices are not sets of individual action.[20] This thesis, however, must be formulated more precisely. A practice *is* a manifold of doings and sayings (basic actions). But a set of doings and sayings constitutes a practice only if its members express an array of understandings, rules, and structure. In expressing this array, they thereby constitute a variety of actions, which are the actions I assume Taylor has in mind in his formulation (e.g., the action of marking a ballot, as opposed to the basic act of scratching a pencil mark on a piece of paper, the performance of which constitutes marking a ballot). As emphasized in my discussion of Oakeshott, these actions are what they are because of the organization of the practice. A practice, consequently, does not embrace a set of actions that possess identities independently of the practice. Its constituent actions are constituted by the practice's understandings, rules, and structure. So a practice is a set of individuals' actions, but not a set of actions defined by reference to individuals alone.

Before discussing intelligibility, I want to focus recent arguments Stephen Turner has brought against the notion of a social practice upon the just elaborated account.[21] I emphasize that his considerations do not, strictly speaking, apply to this account. Turner treats practices primarily as shared mental objects with causal powers. Examples are tacit knowledge, presuppositions, and habits (construed generatively). He skillfully contests the supposition that the best causal explanation of agreement, disagreement, and similarities among actions invokes shared entities of these types. On the present account, however, practices are, in the first place, sets of doings and sayings (as Turner urges – see p. 104, and cf. p. 117). And although doings and sayings compose a practice by virtue of expressing an array of understandings, rules, and teleoaffectivity, these items, as discussed, do not *cause* the doings and sayings involved. This difference neutralizes Turner's broadside, which presupposes an "apparatus" construal of mental phenomena as objectlike causally efficacious entities.[22] Some of his arguments can be reformulated, however, so as to raise issues pertinent to the current account. I will address two such issues.

The first concerns the identification of practices. Turner writes:

> But the only access we have to [practices] is through our own "culture." From the point of view of what we can know about them, or how we can construct them, they are irredeemably cultural facts. We need a starting point *within* culture to recognize something *as* practice. . . . So the very constitution of a practice as an object is tainted by our starting point, which is itself a contingent fact which we can neither understand nor overcome (103)

What someone identifies as a practice depends on his or her position within culture. Turner illustrates this epistemological point through Marcel Mauss's recollection of once being struck by the manner in which his American nurses walked, realizing that he had seen this walk previously in the movies, and then noticing, upon his return to France, that French girls too had taken up this way of perambulating (pp. 20–21). As Turner observes, Mauss's discerning a particular "practice" (in this case, a way of walking), and thus identifying a particular "fact," rests upon his culture-bound expectations and experiences. Convalescing in the same hospital, a person with different expectations and experiences might fail to spot the practice Mauss did and instead name as practice-fact some other feature or pattern of the behavior of American nurses. This shows, Turner writes, that "a 'practice' as such, practice as an object, is inaccessible" (p. 37). Similarly, it might be suggested, specifications of the practices composed by doings and sayings are tied to interpreters' cultural positions. Observers with different cultural trajectories are bound either to represent the same doings and sayings as composing different practices or to collect together different sets of behavior as practices. There is no access, consequently, to the practices that the doings and sayings *really* compose. Or rather, there are no practices in themselves, only culturally relative ones.

As suggested, to which practices a particular behavior belongs rests on the life conditions it expresses; and which conditions these are depends on the behavior, its contexts, and understandings of life conditions. The relevant understandings are mostly those interwoven into the actor's world, into the activities and contexts to which he or she is party. For things can be standing and going for someone only in those ways for which his or her body, activities, contexts, and practices make room. And understandings of these ways are generally interwoven into and carried by the practices involved. Actions, understandings, and practices are thus holistically related; and actors and their we's (cf. Chapter 3) possess most of the understandings against which their behavior expresses particular conditions. A nexus of practices can also, however, make room for conditions of which participants have no understanding. As noted in Chapter 2, people do not have concepts for all the patterns of behavior-in-circumstances traversing their lives.

Whether, consequently, a given person can access (identify) practices "in themselves" is a contingent matter, dependent on her conceptual understandings, knowledge (of the actors and their practices and contexts), and clairvoyance, sensitivity, powers of observation, and so forth. It is no way proscribed *tout court*. Participants in the same practices routinely comprehend both the conditions of fellow participants and the practices they collectively carry on. Outsiders, too, are or can become capable of doing this. Aiding them in this regard is their possession of some if not many of the life condition understandings alive in the participants' worlds. Indeed, human beings share understandings – or at least possess highly similar understandings – of core conditions of life (e.g., expressed in English, believing, wanting, hoping, expecting, being fearful, being sad, trying, and reprimanding). Comprehension and access become more difficult, inter alia, to the extent that practices, and thus certain conditions and understandings, are unique to the people involved.

In other words, that Mauss's powers of discernment were culturally molded shows neither that American nurses did not in fact walk the way he noticed nor that they did not do any of the other things that differently culturally constituted observers might have identified (i.e., observed). Cultural conditioning can certainly occasion distorted perception and interpretation. But all it entails is that what a person notices and how he or she conceives of it is affected by cultural position. Far from automatically thwarting access to practices, the cultural conditionality of perception and thought, for just suggested reasons similar to Vico's, enables access thereto of variable and improvable cogency. Whether, then, in a given case it obstructs comprehension more than underwriting it is an open question. In this context, moreover, it is important to remember that identifications of actions, too, are tied to culture and experience. Turner's reasoning thus implies that the agreements, disagreements, and similarities in action that he takes as givens – and which are that vis-à-vis the explanation of which theories that (culture-conditionally) invoke shared mental object-causes are denigrated as inferior – are in fact not givens, but just as tied to, and "tainted" by, culture as shared mental causes are. Indeed, Turner's logic can be used to argue that all human descriptions whatsoever are "tainted" thusly, thus problematizing the propitiousness of the censure (cf. p. 104).

The second issue concerns conditions of life. Central to Turner's arguments is the question, What is it for the "same" mental object to be in different people? Since doings and sayings belong to a given practice when they express components of the same array of understandings, rules, and teleoaffectivity, this question can be pertinently reformulated as: What is it for people, who perform different behaviors, to share a particular life condition? To simplify the issue, I will focus on practical understanding. Although the considerations relevant here are not entirely the same as those germane to different people either following the "same" rule or expressing elements of the "same" teleoaffective structure, there is considerable overlap. What, then, is it for people to share a particular practical understanding, say, one of questioning and answering?

This question as formulated is equivocal, for the expression "understanding of X" is ambiguous. If two people exchange questions and answers, such that their behavior, barring unusual circumstances, expresses understanding of this activity, the specification of the understanding that their behavior expresses is parasitic upon the activity they are identified as carrying out. Both actors, consequently, display understanding of questioning and answering, and in this sense "share" understanding. Turner would say, however, that this leaves open whether their (own, personal) understandings of the activity are the same. In one sense, they share understanding because it is of the same activity that their doings and sayings express a grasp; in another sense, it remains open whether their understandings of this activity are the same.

Turner's arguments imply that, if understanding this activity is a causally efficacious mental object underlying it, then peoples' understandings differ. For, on his view, the causes of different people carrying on the same activity vary, since behavior arises from bodily causal networks established through activity and experience, which are unique for each individual (see p. 58–59). As discussed, however, the understandings expressed by doings and sayings are not, on the current account, internal states that cause these behaviors. So sameness and difference in understanding cannot consist in the presence of identical or different causally efficacious states of a certain kind in people. The same understanding can be expressed by doings and sayings generated by different (bodily) causes.

What might, on the other hand, suggest that people possess different understandings of the actions they carry out in common are observable differences in their doing so. This consideration seems strongest vis-à-vis language and concepts (p. 74). Myriad small differences among people's use of given words suggests that they understand these words, and the concepts thereby expressed, differently. This line of reasoning raises, however, tangled questions. For instance, when is a use of language "different" and not the "same" use in new, unexpected, or unfamiliar circumstances? (Similarly: When does carrying out an action differently express a different understanding and not the same understanding in conjunction with disparate conditions of other types?) More fundamentally, why do differences in use or activity implicate divergent understandings? Given the indefinite variety of situations of action and the fact that behavior usually expresses more than understanding alone, any understanding will be expressed by varied doings and sayings in both different *and* same or similar circumstances. Behavioral differences are thus compatible with both different and same understandings. When, then, given observable differences, do people nonetheless display the same understanding?

From a Wittgensteinian point of view, I believe, one person uses/ understands language differently from others when his speaking and writing are unintelligible to them (where what is opaque is what is said and not why, matters that are not of course always separate). Similarly, behavioral differences – for example, differences in how one asks/prompts a question or gives (or fails to give) an answer – express divergent understandings when given performances of the activity are unintelligible to others because of their manner of execution. In short, people share an understanding of a word or action when they use that word or carry out the action intelligibly to one another.[23] Not every observable difference in speaking or doing, therefore, expresses a different understanding. Furthermore, although Turner may be right that the causes of performances vary, disparate causal etiologies are compatible with that mutual intelligibility of observably different behaviors that is a worldly manifestation of commonality in people's understanding of the actions these behaviors express. And, to relate this back to the integrity of social practices: Even if there are no collective causal elements responsible for the behaviors that compose a practice, these behaviors nonetheless can and do express the same conditions of life and thus can and do compose a practice.

The Articulation of Intelligibility

I wrote in Chapter 1 that practice theorists champion practices as the central constitutive phenomenon in social life because they view them as the site where understanding is ordered and intelligibility articulated. The current study largely shares this conviction. To prepare the confrontation with alternative practice accounts in Chapter 5, I will now spell out the account of intelligibility and its articulation implicit in the previous two sections. I will develop this account only so far as is necessary to abet the upcoming encounter and not in the depth required to do justice to the huge topic of meaning, significance, and intelligibility.

By "intelligibility" I mean making sense and by "articulation" specification. Intelligibility is articulated through specification of the "what's" of making sense. The "what's" of making sense, moreover, are meanings, or signifieds. So intelligibility is articulated when meanings or signifieds are specified; and a field of intelligibility is articulated when a field of meaning or signification is laid down.[24] When this occurs, we can speak with Heidegger of intelligibility being *gegliedert,* or broken into joints called significations *(Bedeutungen).*[25]

Intelligibility has two basic dimensions: how the world makes sense and which actions make sense. Both dimensions are articulated through the organizations of practices. This means that how things make sense and what makes sense to people to do are molded by these organizations; further, that there are no things making sense and no making sense to do something that are not articulated on the background on practices. To borrow a phrase from later Heidegger: Practice is the house of being (Being and be-ing).[26]

World Intelligibility

World intelligibility is how things make sense; and how things make sense is their meaning. Something's meaning, moreover, is what it is understood to be. This notion of meaning applies to all phenomena encountered in experience and thought. The meaning of an object, for instance, is what the object is understood to be (e.g., a tree, a star, or a house). There is no meaning independently of understanding.

What something is understood to be is expressed in both sayings and doings. Just as the understanding of trees is expressed in the use of the word "tree," what is said about trees, and how people act toward them, a person's understanding of a particular tree is expressed in her calling it a tree, what she says about it, and how she acts toward it (e.g., climbs it, fells it, or admires its foliage). Keeping in mind that understanding is expressed in doings as well as sayings helps hold at bay an overlinguistified conception of intelligibility. Further warding off this widespread but baneful conception is the realization that understanding is acquired through exposure to and the performance of nonverbal as well as verbal behaviors. A person's understanding of trees is acquired not only through exposure to uses of the word "tree" and to speech acts about trees, but also by observing and carrying out such activities as climbing trees, gazing at them, and felling them. Understanding is expressed and acquired in a tightly interwoven nexus of doings and sayings in which neither the doings nor the sayings have priority.

How things make sense is articulated primarily within social practices, for it is within practices that what things are understood to be is established. Within the dispersed practice of X-ing, for instance, it is instituted that people who perform particular doings and sayings in particular contexts are (understood to be) X-ing, attributing or prompting X-ings, or Y-ing or Z-ing in response to X-ings. These matters are "established" in the practice in the sense that people who perform these behaviors have these meanings for anyone who shares the understanding that links and is expressed by the doings and sayings composing the practice. People, moreover, generally make sense to a given person in these ways because of her exposure to and participation in the practice involved, whereby she acquires the understanding establishing these meanings and goes on to perpetuate the practice in her own doings and sayings.

The world intelligibility articulated in an integrative practice is more elaborate and embraces a greater variety of actions than that articulated in a dispersed practice. Within integrative practices, actors not only make sense as Q-ing and as attributing, prompting, and responding to Q-ings, but also as relatedly R-ing, X-ing, and Y-ing (etc.) and as attributing, prompting, and responding to such actions. Moreover, since (1) an actor makes sense to others as Q-ing, R-ing, or X-ing on the basis of her (presumed) other conditions, and (2) the teleoaffective structure of an integrative practice specifies a range of acceptable and correct combinations of conditions, the structure of a practice articulates a person's significance *qua* actor whenever it encompasses the combination of conditions on the basis of which she makes sense to other participants as Q-ing, R-ing, or X-ing. For instance, suppose an office manager sees the public relations chief dialing the telephone. Suppose further that he knows that she intends to give someone else instructions as part of her project of placing an ad in the local newspaper, which she pursues as a step in the recently resolved campaign to improve the firm's public image, which she in turn helps carry out for the sake of fulfilling her job as public relations officer. She then might make sense to him (among other things) as implementing a public relations campaign and doing her job, and do so on the basis of a hierarchy of actions, tasks, projects, beliefs, and ends that is part of the teleological structure of business practices in the firm. On the other hand, if all he knows is that she's always trying to be the best public relations officer she can, and/or in addition that she wants to place an ad in the paper, and/or further that she thinks that Joe is good at soliciting discounts from newspaper people, then she might make sense to him as doing her job, placing an ad, and/or giving Joe instructions. The possibilities are clearly manifold.

The organization of a practice need not and usually does not encompass all conditions on the basis of which what a person does has meaning for other participants. Suppose the manager knows that the public relations chief places an ad in the paper, as opposed to offering to sponsor an exhibit at the local museum, as a favor to the editor who is a friend of hers. Her wanting to do well by her friend, on the basis of which she makes sense to the manager as doing her friend a favor, is not part of the firm's teleological structure. Note that her doing him a favor is also not part of this structure. Just as a practice's teleoaffective structure does not usually encompass all conditions on the basis of which a participant makes sense to other participants as such and such, not all actions constituted by her behavior while she is engaged in the practice will be part of it. At the same time, her doing her friend a favor is part of *other* practices in which she is engaged at the time (e.g., friendship practices); and the structures of those practices *do* encompass the conditions on the basis of which she makes sense to the manager as doing a favor. Thus, as a general rule, actions are components of practices and the meanings people have for others are articulated within those practices. Incidentally, these meanings are not restricted to the actions people are understood as performing, but also encompass the mental and cognitive conditions they are understood to be in (as well as the characters, genders, person types, and so on that rest upon these conditions). Practices articulate this component of meaning, too, by carrying the conceptual understandings of life conditions on the basis of which behaviors-in-circumstances are understood as expressing particular conditions (see the final section in this chapter).

Not only people, but objects (and events) as well acquire meaning within practices. This occurs, most importantly, whenever objects are used in the performance of constituent actions. Teaching, for instance, encompasses writing on blackboards and other surfaces with certain entities, which therewith receive the meaning: things with which to write. Similarly, felling trees involves the use of certain devices, which therewith possess the meaning: things with which to cut down. These meaning are "practical" meanings, and the entities possessing them can be called, following Heidegger, "equipment" (*Zeug*). Like understanding generally, the understanding of equipment is expressed not only in doings (i.e., uses) but also in sayings. People give names to equipment and say of them that they have such and such practical meanings, for instance, that chalk and magic markers are things with which to write and saws and axes things with which to fell. When, moreover, the projects and tasks of a practice hang together and embrace the use of multiple entities, the practice bestows interrelated and partially cross-referencing meanings upon a multiplicity of things. For example, the meaning of a screwdriver, something with which to remove screws, references screws, equipment with which to fasten wood and metal sheets, because the activities of fastening and unfastening hang together as part of building, repairing, and construction practices. An integrative practice, consequently, carries interwoven understandings of interrelated equipment. What's more, since the field of hierarchized actions, tasks, and projects in a practice's organization are those that it is correct and acceptable for participants to carry out, the objects used in the practices enjoy a range of correct and acceptable interrelated and cross-referencing meanings. Just like teleology, practical meaning is normativized within practices.

When a practice, as is usually the case, is carried out in specific settings, the settings are set up to facilitate the efficient and coordinated performance of its constituent actions. The layouts of the settings, as a result, reflect the interwoven meanings that the entities used in these actions possess by virtue of being so used (and talked about). Settings, in other words, are often set up as sites where a given practice or set thereof is to be carried out. When this occurs, their setups derive from the understandings, rules, and teleoaffective structure organizing the practice. The disclosure and layout of equipment within practices also, consequently, exhibits normativity, meaning that things are usually so arranged that they can be easily used in the correct and acceptable ways.[27]

Practices, however, do not bestow practical meanings alone on entities. How people talk about and act toward objects is not exhausted by how they use them. People also observe objects, examine them, measure them, admire them, draw them, and talk about them in numerous ways that do not pertain to use. The meanings that objects thereby come to bear are still established within practices to the extent that the ways of talking and acting in question are components of practices (e.g., those of observation, examination, art and artistic appreciation, science, teaching, celebrating, cooking, and voting). Moreover, something's making sense to someone as Z regularly presupposes her participation in or familiarity with practices in which things are correctly or acceptably understood as Z. This is because people generally assimilate the understandings and intelligibilities of things that are articulated within the practices in which they participate. This holds for the common meanings of people, objects, and events as well as the more systematic and thought-out meanings established in such integrative practices as those of the human and physical sciences.

Practices thus "constitute worlds" in the sense of articulating the intelligibility of nexuses of entities (objects, people, and events), specifying their normativized interrelated meanings. Constituting worlds through meaning does not, of course, bring entities into existence. Instead, practices, by conferring upon entities interrelated meanings coordinate with the actions taken toward them, organize entities into the integrated nexuses that are what reality is and can be for us. As Wittgenstein writes, "Every language-game is based on words and objects being recognized again. We learn with the same inexorability *that this is a chair* as that 2 × 2 = 4" (*OC*, 455, translation corrected and emphasis added).

An important aspect of world constitution is the opening of a space of places at which activities can intelligibly be performed. When a tree is understood as something to climb, for instance, it becomes a place at which climbing is intelligible; similarly, when a platform is understood as something from which to observe the landscape, it becomes a place at which observing is intelligible. In this way, beds are understood as places to sleep, tables as places to eat, and bus stops as places to catch the bus. A place to X is a place where it is understood that X-ing occurs. Insofar, then, that the organizations of practices bestow normativized interrelated meanings upon entities, practices open spaces of interrelated places at which their constituent doings and sayings are correctly and acceptably performed. To use a Heideggerian example,[28] a workshop is an organized nexus of equipment for building, repairing, constructing, and the like that anchors places where the actions performed in these activities can rightly and acceptably take place. The workshop, accordingly, harbors a normativized space of places. Notice that the opening of a space of places is coeval with the establishment of a field of practical meanings; any entity that is a place where X-ing is intelligibly performed is ipso facto intelligibly used in (if not also for) X-ing.[29]

A setting in which a given practice is to occur and which is laid out in accordance with the meanings that objects acquire in the practice's constituent actions anchors a space of places which similarly rests on these meanings. Not only are settings set up to house particular practices, but their setups anchor spaces established by those practices. Practices, consequently, transpire in an objective space that devolves from the material arrangements of objects, while also themselves opening a type of space (the space of places) that differs from and is irreducible to objective space. Emphasizing that practices lay out and course through settings combats the possible misimpression that practices on my account are ethereal constellations of meaning that lack any rootedness in or connection to materiality (more on this in Chapter 6). Emphasizing that they open spaces, furthermore, signals that my earlier characterization of practices as spatiotemporal nexuses of behavior, in an objective sense of "spatiotemporal," does not exhaust the relation of practices to space and time.[30]

Finally, a person's understanding of Z, taken abstractly, is likely to be complex and expressed in a variety of different types of action with regard to it. Z might be something to fell, to admire, to climb, to examine, to prune, to sketch, and so on. Since these understandings are acquired in different practices, the resulting total understanding of Z is a many-colored product. Usually, however, entities are encountered while engaged in a particular practice(s). This means (1) that they will have only those correct and acceptable meanings which are contained in that portion of a person's overall understanding of them that helps organize the practice(s) involved; (2) that they will anchor only that space of places which is coordinate with those meanings; and (3) that people will (usually) talk about and act toward them solely in the correlated right and acceptable ways. Of course, people often suddenly alter which practices they are engaged in, such that entities can abruptly possess different meanings, anchor different spaces of places, and be acted toward differently. Moreover, in a situation where others are carrying out one practice, a person can intentionally understand and act toward entities in ways characteristic of another. Usually, however, people participate steadily in given practices, meaning that they inhabit a world of stably meaningful objects, events, and people.

I wrote in Chapter 3 that a person is "one of us" when she uses language as we do and more broadly acts and speaks intelligibly to us. A "we," consequently, is an open-ended collection of people who behave mutually intelligibly. I also noted that since intelligibility comes in degrees, the boundaries of a we are unstable, shifting, and contingent. We have now seen that practices articulate intelligibility and carry understanding. This implies that people who act and speak mutually intelligibly are, roughly, people who are party to the same field of dispersed and integrative practices. Being one of us, consequently, can be redefined as participating in our practices, where "we" are the people who participate in a particular set of dispersed practices interwoven into integrative ones. This reformulation, be it noted, does not make intelligibility more determinate. Since the organizations of practices are not the exclusive determinant of behavior within them, coparticipants might not be fully or equally intelligible to one another. Similarly, those carrying on practices distinctly different from ours might still slightly or largely be intelligible to us due to similarities and analogies between their practices and those pursued by and otherwise familiar to us.

In any event, someone is one of us when he participates in our shared field of dispersed practices interwoven into integrative ones. Dispersed practices are crucial to a we since they carry understandings of life conditions central to mutual intelligibility. Integrative practices are also key since they carry rules and teleoaffective structures in addition to understandings, and also house contexts and situations in which people act. A relatively isolated society of anthropological lore boasts a single we whose members almost completely share a set of dispersed practices that are broadly integrated into their integrative ones. Such an extreme multicultural spatiotemporal expanse as Southern California, on the other hand, exhibits a variety of we's. Each comprises a set of dispersed practices, to varying extents divergent from but at its core similar to those of the other we's, that are interwoven with integrative practices, many if not most of which are shared with the others. Each of these we's also can have "sub-we's" (e.g., youth) that supplement we practices with more peripheral and exceptional shared activity-manifolds. Alongside smaller we's, finally, the larger space-time unit encompassed by the United States and Canada also harbors one that encompasses many of the inhabitants of these countries and comprises core dispersed practices interwoven with pervasive integrative ones. Notice that a person can belong to multiple we's.

Before considering the second dimension of the articulation of intelligibility, I note that the preceding sketch of world constitution within practices converges with Ernesto Laclau and Chantal Mouffe's description of the articulation of meaning in discourses. Laclau and Mouffe conceptualize discourses as totalities of systematically and interrelatedly meaningful actions, words, and things (they mention the building stones of *PI*, 2). Nonverbal behaviors appear to number among such actions, though the authors are not clear on this point.[31] They analyze meanings, moreover, neo-Saussurianly as devolving from systems of difference and argue that what is for entities to be meaningful (to be what they are) is for them to be "differentially positioned" in such systems (the expression "positions" designates entities *qua* bearers of meaning). They further describe the apportionment of meaning in Lacanian terms as "fixations" of "nodal points" to indicate that meaning is a precarious achievement unremittingly subject to subversion, transformation, and replacement. The term "practices," meanwhile, denotes human activity that arises from extant discourses to transform them and their positions: "Every social practice is . . . – in one of its dimensions – articulatory."[32] In sum, discourses constitute worlds: Nexuses of objects, doings, and sayings articulate systems of difference through which all entities within the discourse receive meanings-identities. Unfortunately (for the present context), Laclau and Mouffe's interests lie in political and not practice theory. Consequently, they do not examine how meanings are fixated and whether the Saussurian figure of systems of difference is compatible with the Wittgensteinian notion of understanding. Moreover, they do not relate their analysis to the second dimension of the articulation of intelligibility, that pertaining to action.

Action Intelligibility

The articulation of action intelligibility is the specification of what makes sense to people to do. What makes sense to people to do, moreover, is "signified" to them as the action to perform. Although people are always able and prepared to do a variety of things, at a given moment they invariably carry out those actions that are signified to them as the ones to perform. I hasten to add that what makes sense to people to do is not the same as what it makes sense to do (i.e., what is rational). It is not even the same as what *seems* rational to actors. Signifying can diverge from the counsels of rationality, both impersonal and ostensible. This will become clearer as the discussion progresses.

It will be useful to begin with the Wittgensteinian idea developed in Chapter 2, that doings and sayings express mental and cognitive conditions. As we saw, although mental conditions such as being in pain, imagining, being joyful, and fearing are continuously expressed, this is not true of cognitive conditions such as intending, believing, desiring, hoping, and knowing. The cognitive subset of the conditions expressed by a given doing or saying, consequently, is not continuously expressed while the actor is in them. Furthermore, since cognitive conditions are not usually manifested but only signified in behavior, it is rarely possible to read off directly from behavior which cognitive conditions it expresses. As G. E. M. Anscombe brilliantly explored, however, one way cognitive conditions and orders thereof can be revealed is through the answers someone gives to a series of why questions concerning his behavior.[33] This holds even when a later report of a condition is the key element of the context by virtue of which what the reporter's earlier behavior expressed was the reported condition (e.g., *PI*, 487, 682; *Z*, 14).

In Anscombe's example, a person is pumping water into the cistern that supplies the drinking water of a house. When asked why, he replies that he's replenishing the house's water supply. This response reveals his desire to replenish the supply, his intention to do so, and his belief that pumping is the way to accomplish this. Suppose a further why question – for example, Why do you want to replenish the supply? – prompts the reply: to poison the people up there. This response reveals further conditions (including actions). In this way, an extended series of why (and how questions) can reveal the inter-connected cognitive conditions expressed by his operating the pump handle. Some such order as this, Anscombe writes, "is there when-ever actions are done with intentions."[34] It is important to emphasize that series of why questions primarily reveal the teleological dimen-sion, or ends-orientation, of action. The array of cognitive conditions that any such series discloses houses the teleology that governs activ-ity: They lay out what the actor saw in an action. This implies that whenever teleology is absent (e.g., in many instances of crying and laughing), why questions (about what one is up to) will fail to reveal the conditions that are expressed in the behavior and determinative of action intelligibility. In these cases, there is nothing that the actor saw in his or her action. Other devices or contexts, consequently, must be summoned to apprehend the conditions expressed in and explanative of activity.

I emphasize, further, that even when a person simply reacts, that is, acts without the intervention of explicit thinking, she can be in a collection of life conditions expressed in that reaction (she can also of course be in further conditions that this reaction does not express). Dreyfus argues that a person is not in any cognitive conditions what-soever when her behavior is a flow of reactions.[35] She is in such conditions as intending, believing, and desiring only in breakdown situations, when reactions are interrupted and deliberate attention and thinking intervene. Dreyfus claims this, however, because he construes intentions, beliefs, and desires as "mental states," that is, states that (1) are explicitly directed toward something and (2) cause or guide behavior.[36] On the Wittgensteinian account presented here, by contrast, cognitive conditions are not explicitly directed states with causal powers.

Of the different categories of life condition, only the objects of states of consciousness are inherently explicit. Because, consequently, a person is more or less always attending to something (which may not, however, be that toward which she acts), she is invariably, even when acting unreflectively, in at least one condition whose object is explicit to her. This need not, however, hold of the objects of her emotions, beliefs, and intentions; and in the normal course of life these objects are unthematic. What's more, my upcoming claim, that the conditions expressed by actions (including reactions) lay out the "reasons" why they are performed, does not imply that these condi-tions cause actions, even reflective ones. Thus, in sum, cognitive conditions are not explicitly directed states with causal powers. They are not, consequently, incompatible with reactivity. Reactions express beliefs, intentions, and desires (as well as emotions and states of consciousness); and these conditions explicate the reactions without causing them. While Dreyfus is importantly right to insist that all thematic intentionality occurs on a background of transparent coping[37] (i.e., spontaneous reaction), it is equally true that coping expresses actors' unthematic conditions of life.

Oakeshott's account of action nicely acknowledges this fact while still succumbing to the pervasive temptation to overintellectualize action. As explained in the second section of this chapter, what a person does, on Oakeshott's analysis, is determined by her understanding of her situation, a thereby resulting wish, and her motives, or sentiments. He writes more expansively that whenever a person's behavior counts as action (or, rather, conduct *inter homines*), the person evinces reflective consciousness, intelligence, understanding, deliberation, imagination, wishing, choice, belief, seeking, and sentiment. These phenomena are not

"goings-on" which may or may not precede actions or utterances and which are either consciously engaged in or are absent. . . . [I]f human conduct [evinces] reflective consciousness it does not follow that what is done or said must be done or said reflectively or self-consciously in order to qualify as conduct.[38]

These phenomena are, instead, "postulates" of action, that is, conditions of action in terms of which it can be understood (p. 42). This means, I think, that whenever behavior counts as action, an entire apparatus of concepts – of reflective consciousness, intelligence, understanding, deliberation, imagination, choice, and the like – applies to it. Oakeshott's analysis thus parallels the account offered here. Such conditions as imagining, choosing, understanding, deliberation, and reflection are not explicitly directed states that either are engaged in consciously or not at all. Unreflective action does not exclude an actor's being in and expressing such conditions and thus the application to him of their concepts.

Despite his practical sensibility,[39] Oakeshott still misconceives action in contending that a specific and, moreover, elaborate apparatus of such concepts (postulates) applies to all actions. Acting, he writes, "is diagnosing a situation, recognizing it as an invitation to act, imagining a satisfaction, making a choice, performing an action, and encountering an outcome" (p. 34). This same organized network of concepts applies to any action. In my view, on the other hand, which conditions behavior expresses is a contingent and variable matter. Nothing guarantees that a particular well-defined and organized network of types of condition (even a network of cognitive conditions spelling out a teleological hierarchy) is there whenever people act. Oakeshott, consequently, oversystematizes action and in this way overintellectualizes it.

In any event, the particular conditions that *are* contingently and de facto expressed in behavior thereby articulate the intelligibility of the behavior involved. Collectively, they lay out how things stood and were going such that a particular action was for the actor the thing to do. They thus spell out *why* he performed it. For instance, suppose one person's reprimand of another expressed, inter alia, the beliefs that the second person acted rashly and that such behavior is detrimental to group success, such that the reprimand was delivered in order to correct this behavior for the sake of maintaining group prospects. In this case, the expressed conditions house (part of) the teleology, the orientation toward ends, that governed the delivery of the admonishment. Suppose, on the other hand, that the reprimanding did not express these beliefs, but instead, inter alia, the snarly mood the actor was placed in by the nondelivery of that morning's paper together with the unusually heavy traffic during the morning commute. In this case, the reprimand was delivered (partly) because of how things were going for the actor at the time. It is not explained teleologically, by something the actor saw in it, but instead affectively, by how things mattered. Note that action intelligibility pertains not only to intentional actions (teleologically governed actions to which Anscombe's series pertain), but also to such unintentional actions as crying that are largely governed affectively.

Since an ordered set of an actor's conditions spells out how things stood and were going such that a particular action was the one to perform, the set cannot merely be expressed in the action, but must somehow also relate to its do-ing. One possible connection is that the set causes the action. As explained, however, life conditions do not cause the actions that express and are thereby explained by them. The set also usually does not connect with actions by passing through consciousness. Most behaviors are unreflective reactions ungoverned by conscious thought. This means that most of the explanatory conditions a person is in are not states of consciousness; and that an actor usually does not perform an action after consciously considering that he wants a, expects b, sees c, and therefore should X.

Here is where the notion of signifying comes into play. At any given moment, a person is in a number of conditions; that is, things stand and are going for him in certain ways. Part of what it is for things to stand and be going in these ways is that certain actions and not others make sense to him to perform at that moment in the current situation. These actions are *signified* as the ones now to perform, even when as is usual neither the signifying nor the action signified is explicitly entertained. The signifying instead focuses and channels the flow of unreflective action onto the performance of particular actions. And it accomplishes this (cf. Chapter 2) by singling out an action – what is now to be done – that can be carried out without further ado by performances that are part of a person's spontaneous bodily repertoire. As what is now to be done, signifying specifies an action that can be performed immediately. Moreover, since the transition from signifying to performance is automatic, explicit thought need not intervene in the flow of and at the joints between performances. Explanatory orders of conditions thus connect with the do-ing of actions by being such that, to actors, there are signified actions now to be performed (including speech acts) that they can and do automatically carry out through the performance of bodily doings and sayings.

The notion of signifying (*bedeuten*) hails from Heidegger. On his analysis, what a person does is structured by *Verstehen* (understanding) and *Befindlichkeit* (attunement). Understanding houses the teleological component of the structuring of action, while attunement is the locus of its affective component. Together, they lay out why, at a given moment in a particular situation, an actor performs a particular action.[40] As sketched in Chapter 2, the teleological dimension of signifying consists of "signifying chains" (my expression), which stretch from possibilities of existence for the sake of which (*worum willen*) someone lives to particular actions that are signified as what to do at particular moments in particular situations for the sake of those possibilities. Such chains pass from for-the-sake-of's through series of toward-which's (*Dazu*) and in-order-to's (*Wozu*) before signifying specific actions. Stated in now more familiar language, the idea is that an end, say, winning Teresa, signifies (to Michael) a particular project, say, buying her flowers, which in turn signifies a particular task, say, driving to the nearest florist, which itself signifies to him the action of getting the keys, which in turn signifies looking for them – so he stands up and begins looking about the room, these being doings in his bodily repertoire, the performance of which in that situation constitutes looking for the keys; and looking for the keys being signified as what to do now in the current situation for the sake of winning Teresa. Note that a signifying chain is not a temporal process, but the structure of an occurrence (signifying). Note further that signifying does not always possess a structure as elaborate as in the preceding example. A person might, for instance, order white chocolate for dessert simply because she likes it. All signifying chains, however, terminate in a "for the sake of" (an end), for ends are simply the de facto conclusions of chains, whatever they are, however particular they might be (e.g., enjoying white chocolate), and however long or short the chain. It must be kept in mind, furthermore, that ends need not be consciously entertained.

The teleological dimension of signifying bears an obvious and sharp resemblance to a string of answers given to a series of Anscombe's questions. The conditions of life revealed by, and maybe also mentioned, in those answers do not, however, appear in Heidegger's description. To analyze the structure of signifying, he instead uses such terms of art as "for the sake of," "toward-which," and "in-order-to," which can be rendered in more familiar language as end, project/task, and purpose. Heidegger is forced to adopt terms of art because, to state matters perfunctorily, his aim is to describe the structure of existence; and terms for life conditions are ill suited for this task since they are regularly used in social life for *other* purposes (such as answering questions and explaining oneself and others). The two sets of vocabulary can, however, be related. For Heidegger's language captures the difference that the conditions designated by common mental locutions make to the specification of what to do in the flow of behavior. It articulates the bearing that a person's conditions have on what that person does. For instance, desiring or wanting X can be construed as the project or end of acquiring, doing, or being X helping to specify action. Believing Y, moreover, can be construed as the state of affairs, that Y, helping to determine that a particular action made sense to someone for the sake of some end.

The second dimension of signifying is attunement: things mattering (*angehen*) to people. Things mattering is people's being in particular moods and emotions or having particular feelings, affects, and passions. How things matter omnipresently structures the stream of behavior. It usually accomplishes this by affecting what is teleologically signified as the thing to do. It achieves this, in turn, by affecting either which possibilities of existence a person is out to realize or what makes sense to him to do for the sake of these possibilities. Since most people act teleologically most of the time, mattering primarily determines doing by shaping people's orientation toward ends. As suggested, however, mattering can structure activity independently of an actor's ends and thereby overturn the teleological character of action. For instance, someone might compliment an acquaintance for a past achievement upon winning an award, not for the sake of anything, but simply out of an elated and magnanimous countenance toward things. As mentioned, moreover, crying (usually, and in many cases laughing too) is governed only affectively. Not all actions, consequently, are governed by teleological orders.

Teleoaffectivity governs action by shaping what is signified to an actor to do. This means that the thing to do either derives from the actor's ends and projects, given particular states of affairs and how things matter, or reflects simply how things matter in a given situation. I noted earlier that the conditions of life expressed in an action lay out how things stood and were going such that it was the thing to do. It follows, as I have been suggesting, that the conditions expressed in an action house the teleology and affectivity governing it. As indicated, a teleoaffective order is in fact an order of life conditions.

When, then, do practices help articulate action intelligibility? Integrative practices accomplish this when the teleoaffectivities governing behavior conform to the explicit rules and teleoaffective structures that organize these practices. Teleoaffectivities heed rules when what is enjoined in a rule helps structure signifying because it is enjoined. They conform with teleoaffective structure when they either instantiate a correct teleoaffective order because it is correct or instantiate an acceptable order and it is also true that, had the order been unacceptable, it would not have been realized. (As suggested, teleoaffective structures establish, inter alia, a field of correct and acceptable ends, a selection of acceptable or correct projects to pursue for the sake of those ends, a variety of acceptable or correct tasks to carry out as part of those projects, a range of acceptable or correct ways of using objects, and a variety of acceptable and even correct emotions, feelings, and passions.) The more totalitarian the practice (e.g, some military practices), the more its organization specifies correct (as opposed to merely acceptable) teleologies, behaviors, and even affectivities, and the greater the extent to which it determines the structure of signifying. The less totalitarian the practice, the more its organization establishes acceptable alternatives, and the more individualized and less practice based are the teleologies and affectivities that determine what is signified to people to do.

I emphasize that the teleoaffective order that governs an action might not fully conform to the organization of any given integrative practice the actor is carrying on at the time. Components of this order might, for instance, simply be irrelevant to a given practice and its organization. When an American farmer blesses his fields before planting grain, the signifying of this behavior is partly determined by beliefs that are irrelevant to American farming practices. In addition, people sometimes disregard or actively contravene the teleoaffective structures of practices. The farmer might also, for instance, regularly burn his fields to improve their productivity.

Dispersed practices, meanwhile, shape signifying to a smaller degree than integrative ones do. The second component of the understanding that organizes such a practice, the ability to identify and recognize X-ings, enables participants to comprehend one another (and equipment), whereas the third component, the ability to prompt or respond to X-ings, enables correct and acceptable promptings of and responses to this action. The first of these know-hows shapes signifying whenever what makes sense rests on this comprehension. The second does so either when one of the prescribed ways of prompting or responding is signified in part because it is prescribed or, more weakly, when one of the acceptable promptings or responses is signified and it is also true that, had it been unacceptable, this would not have occurred. Dispersed practices, however, do not otherwise figure the signifying that governs their constituent actions. So their organizations usually do not fully disclose why these actions are performed. (This is why dispersed practices can meander through integrative ones.) Similar remarks apply to the understandings carried in integrative practices. Finally, the first component of the understanding that organizes a dispersed practice, being able to X, is also pertinent to signifying. For a participant in the coordinate practice to be able to X is for her bodily repertoire to be such that she can X through the spontaneous performance of actions in the repertoire. Since she can X without further ado, signifying does not need to specify what to do in order to X. It forthwith yields, consequently, to performance whenever X-ing is signified. In this way, understanding, in the form of bodily repertoires, molds action intelligibility.

Although the teleologies and affectivities that govern an action might not be fully shaped by any practice the actor carries on when performing it, what makes sense to people to do is still nurtured within the wider ambit of practices. For the teleoaffective orders that govern actions are usually molded without remainder by the combinations of practices actors carry on at the time. And those components of governing teleoaffectivities that are not encompassed by the organizations of concurrently carried-on practices nonetheless develop within, and are shaped by participation in, these and other practices. Human life generally transpires within social practices: Actions are always carried out while participating in practice(s), and the shaping of mind that accompanies someone's presence in and interactions with the world transpires as she participates in and encounters practices. Hence, the articulation of action intelligibility occurs *within* social practices, regardless to the extent to which it is shaped *by* the organizations of practices. Although practice organizations are not mirrored in the "structure" (organized contents) of mind/action, practices remain the central and omnipresent formative context of structured human existence.

I conclude this section by pointing out that, since practices are inherently social entities, so, too, are the world and action intelligibilities they articulate. Of course, entities make sense however they do *to* individuals, just as whatever is signified as the thing to do is always signified *to* an individual. Since, however, practices articulate how people understand things and shape what makes sense to them to do, intelligibility is a social determination.

How things make sense to someone and what is signified to her to do rest squarely on the social practices in which she participates. To the extent, then, that individuals participate in the same practices, world and action intelligibility are articulated for them alike, differences always accompanied by commonalities and tied to the possibilities opened for people within practices. More momentous and conspicuous disparities and conflicts arise from the different mixes of practices people carry on. When the differences become sufficiently large, or line up along specific fault lines (as are sometimes defined by blocklike agglomerations of linguistic, family, religious, and celebration practices), people find themselves confronted with "cultural" or "societal" divisions in understanding and intelligibility.

Intelligibility and Language

Via formulation, expression, and enforcement, language plays a critical role in the articulation of intelligibility. Words are used to formulate rules, to express understandings, teleologies, and affectivities, and to enforce the normativity pertaining to these matters. Many writers, in fact – for example, later Heidegger and most of those working under his impress – aver that the limits of language are the limits of intelligibility. The current section combats this intuition and argues that the limits of intelligibility are charted in behavior out beyond the limits of what can be said. Instead of pursuing a general argument to this end, I will first describe an eminent social ontological version of the contested thesis and then offer considerations undermining it.

In a number of writings, but especially in "Interpretation and the Sciences of Man," Charles Taylor suggests that the range of "meanings" constitutive of a social practice is tied to the array of vocabulary used within it. The meanings Taylor has in mind are primarily life conditions people can be understood as being in, above all actions they can be understood as performing. Under "meaning" he also includes the significances that situations can have for people, since the significance of a situation is tied to people's conditions. That a situation is terrifying, for instance, is bound up with someone's being in fear; that what she feels is shame, moreover, is coordinate with her situation being humiliating (p. 23). A social practice thus possesses a structure of meanings that people and situations within the practice can have for its participants. Taylor calls these meanings "intersubjective meanings." He contends, further, that a structure of meaning does not exist independently of people's interpretation of them. Because, therefore, interpretations are linguistic articulations, a practice has no structure of meaning independent of the language with which its participants interpret situations and people within it (including themselves). For instance, situations can't be humiliating or terrifying and people can't be fearful or ashamed independently of their possessing a language with which they can describe situations and themselves thusly. So the range of meanings that people and situations can have for participants is tied to the language in which their self-interpretations can be couched.

This latter claim is further explicated through the notion of a semantic field. Meanings, Taylor claims neo-Saussurianly, are defined within fields of contrast. A meaning is the meaning it is by not being a range of other meanings. Deference, for example, is what it is by contrast to respect, cringe, mild mocking, irony, insolence, provocation, and downright rudeness (p. 22). Any meaning thus presupposes a field of meanings in which it receives its identity. Taylor further claims that a social practice is carved up into a range of possible actions. Negotiation practices, for example, encompass entering into negotiation, breaking off negotiation, offering to negotiate, negotiating in good (bad) faith, concluding negotiations, making a new offer, and so on (p. 32). Once again, that a piece of behavior is one of these actions is not independent of people's linguistically articulated interpretation of it as such, thus not independent of its bearing a certain description for them. In Taylor's words, a behavior's identity as one of these actions represents the "application of a language of social life to it" (p. 32). So, because this meaning (like all meanings) presupposes a field of meaning, it also presupposes a range of vocabulary marking the field. "The semantic 'space' of this range of social activity is carved up in a certain way, by a certain set of distinctions which our vocabulary marks; and the shape and nature of these distinctions is the nature of our language in this area" (p. 32). The range of possible actions in a practice is marked, consequently, by the vocabulary at work in the practice.

In sum, a practice embraces a range of meanings that cannot exist without being marked in the language utilized in the descriptions that the practice's behaviors and situations have for participants. So "the distinction between social reality and the language of description of that reality is an artificial one" (p. 34). The limits of the intelligibility articulated in a practice are one with the limits of the language spoken there: People and situations can have only those meanings within the practice that are marked in its language. They can have no intelligibility inexpressible in that language.

My critique of this position aims only at its edges. Most of the meanings things can have for people are indeed marked in the language they can use to speak of them. The language in which they discuss people and behavior similarly registers the orders of teleoaffectivity that structure action intelligibility. I also affirm that world and action intelligibility would probably be at best rudimentary in the absence of a language in which to express and formulate them. It is still, however, worth resisting the assimilation of intelligibility to language effected in Taylor's identification of the distinctions and possibilities of intelligibility with the distinctions and possibilities marked in language. Intelligibility is ultimately and (one presumes) originally a practical phenomenon that is not entirely recouped in language.

Before showing this, it is important to point out that discussion of this issue is bedeviled by the fact that language is the medium in which analysis is conducted. Whenever we think about practices and actions, we do so in language and thereby attribute to practices and actions what is formulated in the language employed. This trivial fact can have grievous consequences in certain situations. For instance, I wrote earlier that understanding is expressed in both what people say and how they act. If it is asked, What understanding do sayings and doings jointly signify?, this can be thought about only with the resources furnished by language. So it is naturally tempting to take as the content of this understanding something language enables us to say. But this is in effect to assimilate this understanding to what is and can be said in language. To resist the assimilation, one must resign oneself to asserting no more than that there is something to understanding other than these formulations, or that understanding is not completely formulable, or that understanding is also expressed in what people do. Such statements do not attempt to say what is unformulable in words, but instead suggest abandoning the attempt and that the best one can do to "familiarize" oneself with this understanding is to scrutinize, learn, and participate in practices in order to acquire it. Dedicating oneself to gaining this sort of understanding would mean, of course, the end of analysis. Even after a period of learning and doing, moreover, the same quandary would reassert itself upon the resumption of reflective inquiry. The moral of this is simply that the only way of showing that the limits of intelligibility are broader than those of language is to show that not all intelligibility can be formulated in language. There is no saying what more there is.

We have already seen that speakers' understandings of many words, including those for conditions of life, cannot be exhaustively formulated in words. The living understanding of these conditions is richer and suppler than what of it can be captured in linguistic formulations. What it means to be in a particular such condition, consequently, cannot be fully spelled out; that is to say, what someone, who is understood to be in that condition, is understood as being cannot be fully formulated. The intelligibility of persons (and of world more generally), as a result, cannot be completely laid out in words.[41] I tried in the previous chapter to signal this richness by suggesting that what is grasped in understanding the (concept of) a life condition is the configurations of relevancy and lines of intelligibility that figure and bind together behaviors-in-circumstances as expressions of that condition. Talk of configurations and lines is a way of repudiating determinate formulations of what is grasped in understanding these conditions. This way of describing understanding also reveals that (the use of) a condition-of-life locution masks behind the veneer of a single signifier designating a single condition the particularities of the expressive bearings of a wealth of different behaviors-in-circumstances. The use of the word thereby fails to register these particularities, which are revealed only (if at all) by other reactions to the behavior involved.[42]

Also obscured by this telescoping effect is the phenomenal wealth and variability of the contextualized behavior involved. That people sometimes attribute life conditions to one another on the basis of, in Wittgenstein's words, "imponderable evidence" (e.g., *PI*, p. 228) – something indescribable in their behavior, tone of voice, or facial expressions – further reveals this occlusion. The existence of such evidence shows not only that people are attuned to features of behavior for which they lack words, but also that these features are poorly registered in language (only as well as the descriptions furnished). Understanding of these features is nonetheless evidenced in the ability to attribute conditions; and it is further expressed wordlessly in the ways people react to one another. Dreyfus points out similarly that

[i]n complex domains one does not have words for the subtle actions one performs and the subtle significations one Articulates in performing them. A surgeon does not have words for all the ways he cuts, or a chess master for all the patterns he can tell apart and the types of moves he makes in response.[43]

Indeed, because a vast range of know-how's, from knowing how to ride a bicycle to knowing how to make a weather forecast,[44] are unformulable, the practical meanings that entities enjoy in the activities underwritten by these know-how's similarly cannot be adequately expressed in words.

Hence, people's understanding of one another, as manifested in (1) their nonverbal reactions to one another, (2) what they say about others on the basis of behavior, and (3) what they manage to say about the features of behavior on the basis of which they say these things, is richer than the distinctions and categories marked in language. Of course, we can't say what the greater richness is; what more there is cannot be formulated because whatever is said will be said in the categories of language. But it is still given to us to know that understanding is richer than the possibilities of linguistic expression. We note the impossibility of adequately formulating living understandings of life conditions, the imponderable evidence on the basis of which people attribute these conditions, and the subtle differences in how they nonverbally react to others.

Again, none of this denies the crucial role language plays in articulating intelligibility. It does, however, contravene the assimilation of the limits of intelligibility to the limits of language. The meanings that humans and situations have for people are not and cannot be exhaustively captured by what they say about these things, a fortiori by the vocabulary employed in discussing them. So the "semantic fields" of practices are not adequately charted by the languages employed within them. Of course, causally speaking, as noted, meaning transcends language only because language pervades practices. Anything more than the simplest intelligibility most likely requires the presence of linguistic articulation. Language thus possesses the remarkable property of helping to sustain a complex field of intelligibility that extends beyond its reach.

Given Taylor's thesis that the life condition language employed in a practice helps chart that practice's semantic field, I should point out that common life condition locutions are not generally associated with specific integrative practices. Such locutions, especially those for mental and cognitive conditions, are instead carried in the dispersed practices that Wittgenstein calls language-games with particular words. These practices hang together with one another and, in meandering through different integrative practices, tie together different domains of practice. Common terms for conditions thus form a general depository that integrative practices can draw upon. They signal and mark a general background of possible conditions for participants in these practices. Integrative practices do nurture to varying extents their own additional possible life conditions, especially actions; and these annexes are sometimes marked by distinct vocabularies. But the action terms involved are not (at least at first) common ones. What's more, the additional mental and cognitive conditions harbored in integrative practices are for the most part further contents of conditions, not new types of conditions (e.g., what is believed versus believing). They are not, therefore, marked by peculiar life condition vocabularies.

Despite these criticisms, Taylor is crucially correct that the linguistic articulation of intelligibility (however exhaustive it may or may not be) further discloses the distinctness of practices from individuals. The meanings and possibilities marked by the terms of a public language are "not just in the minds of the actors but are out there in the practices themselves" (p. 36). For linguistic terms have meaning only in use, and use is a feature of ongoing practices. So the meanings of terms, and therewith the possibilities they mark, are out there in these spatiotemporally evolving entities. Meanings and possibilities are also, as Taylor acknowledges, grasped by individuals. But there are meanings for individuals to grasp only because practices sustain them. The ranges of meaning and possibility marked by language are carried in the practices in which it is used, regardless of how few, how many, or exactly who grasps them.

How Practices Constitute Mind/Action

Conditions of life are not practices. Nor are persons subsets of practices. Life conditions are aspects of how things stand and are going for people that are expressed in doings and saying. Which conditions are expressed by a given doing or saying depends on the behavior involved, the contexts in which it is performed, and understandings of life conditions. Persons, meanwhile, are most centrally that to which conditions of life are attributed. When conditions are expressed in doings and saying, there is *eo ipso* someone in them to whom they may be ascribed.

Practices help constitute mind/action in three ways. First, people acquire the abilities to carry out the doings and sayings that express conditions of life – at least at the current stage of knowledge and technology – through exposure to and above all participation in the manifolds of linked doings and sayings that compose practices. As discussed in the previous chapter, expressive bodies are social products fashioned by the folding and unfolding of human bodily activity into these manifolds.

Second, practices represent one extremely important type of context determining which conditions are expressed by particular behaviors. That behavior expresses one particular condition rather than another usually depends on which practice(s) the actor is carrying out when performing it. One and the same wave of the arm, for instance, can signify the intention to turn or the desire for a teammate to position himself closer depending on whether the actor is participating in driving or football practices. It is because action usually conforms to the understandings, rules, and teleoaffective structures that organize the practices the actor is currently carrying out (see the subsection on action intelligibility) that these practices centrally determine which conditions are expressed by his behavior.

As discussed, furthermore, understanding a life condition is expressed in the use of the locution for that condition together with actions taken toward persons identified as in it. This understanding is usually carried in a dispersed practice of the sort that Wittgenstein calls language-game with the word "x." Since this practice hangs together with other such practices and wanders through diverse integrative ones, it is not unpropitious to say that a nexus of dispersed and integrative practices carries this understanding. Integrative practices also carry understandings of the activities they establish. Thus, third, practices constitute mind/action because the understandings, on the background of which behaviors express particular life conditions, are carried in practices. In particular, the background of understandings against which a given behavior expresses particular conditions is carried primarily by the nexus of dispersed practices within integrative ones in which the actor lives. Like the development of an expressive body, moreover, a person acquires understandings of these conditions by exposure to and participation in these practices where he or she encounters doings and sayings expressive of them.

Hence, people's conditions depend, inter alia, on (1) the particular practices they carry out at a given moment and (2) the wider nexus of practices they enact and encounter that nurtures bodily abilities and carries understandings of life conditions. As I wrote in Chapter 2, the contexts constitutive of expression are shaped within practices. It is here that intelligible as well as paradigmatic patterns of behavior, combinations of conditions, and situational relevancies are established and come to inform behavior; and also here that understandings of the conditions expressed by ever-changing patterns and lines of relevancy are laid down and acquired.

Even fields of possible conditions are established within practices. Practices carry the techniques for using the range of terms that designate such conditions; the variety of nuanced reactions to behavior, marking understandings of others richer than what can be formulated in language, is molded within practices; the behaviors and patterns thereof characteristic of particular conditions are laid down and assimilated within practices; and the range of situations that people can find themselves in is established mostly within particular manifolds of activity.

5

Dimensions of Practice Theory

Waves of theory have rolled in in recent decades to uproot the rigidity of the venerable opposition between individual and society, as well as the viscous front between subject- and object-oriented schools of thought. The previous chapter, in pushing beyond the ideas on mind/action and social constitution extractable from Wittgenstein's texts, initiated engagement with one prominent such strand of thought, "practice theory." The present chapter documents the broad convergences and ultimate divergences between my account and the two grandest and most influential analyses of practice, those of Bourdieu and Giddens.[1] Its narration will foster deeper understandings of both the nature of practice and key themes that link practice theories.

One such theme is the consolidation of two of the three notions of practice distinguished early in the previous chapter: practice as spatiotemporal array of linked behaviors and practice as the do-ing of behavior. Animating this consolidation are, first, the realization that practices *qua* spatiotemporal manifolds only exist through individuals performing the doings and sayings that compose them, and, second, the resulting intuition that these practices and behaviors must somehow be structured alike. The centrality of this consolidation in the analyses of Bourdieu, Giddens, and the previous chapter is evidenced in the prominence they accord to elaborating a structure that at once organizes social practices and governs their constituent performances. Much of the current chapter concerns their divergent accounts of this unifying structure.

A second commonality is the diffuse background influence that Wittgenstein exerts upon the attempts of practice theories to steer elsewhere than back and forth between individual and society and subject and object. Many practice theorists (e.g., Giddens, Taylor, Lyotard, and Laclau and Mouffe) are at least vaguely oriented and inspired by Wittgenstein's remarks on language, meaning, and activity. The extent to which they draw explicitly on his texts varies, of course, just as their understandings of specific Wittgensteinian themes (e.g., rules) diverge. I nonetheless mention this commonality in part because it suggests that something is right and something else wrong about Giddens's remark that Wittgensteinian philosophy has failed to theorize the social institutions it presupposes (*CP*, pp. 49–50). Although this is true of most analyses in Anglo-American philosophy, it cannot be asserted of the wider, though more nebulous cloud of practice analyses to which Giddens's own work belongs. I believe that the notions of practice developed in the current work better accord with Wittgenstein's remarks than do those advanced by other practice theorists. Still, their accounts of social phenomena are impacted in fundamental ways by his thought. So Wittgensteinian "approaches," widely defined, have and continue to say something about institutions.

Discursive Practices

I will preface my examination of Bourdieu and Giddens with a word about Lyotard's appropriation of Wittgenstein's notion of a language-game. His interpretation is so radically wrong that it points by negation to a key aspect of behavior/practice around which other practice theories revolve.

Language-games, in Lyotard's hands, are collections of phrases (*phrases*). Phrases, meanwhile, are more or less the same as speech acts. They are not, consequently, linguistic utterances or statements alone, but any behavior that says something about something. It must be emphasized, however, that linguistic utterances/statements are Lyotard's main interest and focus. This is revealed by the numerous places where he propounds a general thesis about phrases by asserting something about language.[2] In any case, Lyotard countenances sayings alone as the acts constitutive of language-games. Doings are more or less utterly neglected. Language-games, consequently, are purely discursive practices. This remains the case in *The Differend*, where Lyotard divides the notion of a language-game into those of a regime and a genre. Regimes are more or less the same as types of statement or speech act (e.g., cognitive, prescriptive, interrogative, exclamative, and performative [sect. 178], and reasoning, knowing, describing, recounting, questioning, showing, and ordering [p. xii]), whereas genres (of discourse) represent different modes of linking phrases onto one another (e.g., tragic, technical, dialogical, logical, ethical, speculative, dialectical, narrative, erotic, and rhetorical). Although this distinction multiplies and thus deepens his analysis of the pragmatic dimensions of discursive practices, it sustains his exclusionary focus on discursiveness.

To maintain that there are well-defined social entities consisting in sayings alone is egregious enough. Lyotard compounds his misdirectedness by more or less equating social life with nexuses of language-games. When he first raises the question of the relation between language-games (i.e., discursive practices) and sociality, he wisely declines to equate them, instead maintaining simply that language-games are the "minimal relation necessary for society to exist."[3] This reserve evaporates, however, in subsequent formulations, for example, the oft-quoted phrase, "This 'atomization' of the social into flexible networks of language games . . ."[4] In *Just Gaming*, language-games are unequivocally equated with "the social bond" (the bond relating individuals): The statements that constitute language-games "position" individuals vis-à-vis one another both as addressors and addressees of phrases and in what is said (sense) about something (referent) by the phrases involved.[5] Sociality, consequently, consists in the relatedness among people established by the pragmatics of statements/utterances. Lyotard has no qualms writing that the self lies at the nodal points of plural communication systems (but: no other types), and that institutions add further constraints to what should and should not be said (but: not also done) over beyond those imposed by the rules governing language-games.[6]

The equation of social practices with discursive practices is not satisfactory as either interpretation of Wittgenstein or independent social analysis. Wittgenstein clearly embeds sayings within matrixes of sayings and doings. Lyotard cannot reply that the doings "depend" on the sayings in some fundamental way in which the reverse is not the case, for example, that the doings are always responses to sayings (as carrying out an order responds to an order), but not vice versa. Not only is it difficult to see how the interdependencies among the doings and sayings composing a social practice neatly sort out to give systematic and fundamental priority to sayings; but in a number of well-known passages Wittgenstein writes that action (not thought) underlies language (e.g., *OC*, 475; *Z*, 391). In Wittgenstein, as Giddens remarks, the "spacings" by virtue of which words acquire meaning are opened in the organization and progression of social practices (*CP*, p. 36), in the manifold of already existing ways of behaving that are ultimately rooted in brute factual modes of reaction.

Depicting practices as purely discursive is no more propitious as independent social analysis. The linked wordless doings of factory workers help constitute industrial practices, even though they are neither discursive behaviors (usually) nor moments of purely communicative practices. Similarly, a child's playing ball by herself in the street to pass the time before her parents return from work is obviously as much a part of recreational (or play) practices as is the game of twenty questions she plays with her parents when they finally arrive home. Examples can be multiplied indefinitely. In fact, the irrepressibility of doings as components of social practices occasionally surfaces in Lyotard's texts. Scattered among the myriad examples of sayings linking onto sayings are rare cases of doings doing so, for instance, a salesperson linking onto the "tragic" phrase, "How these vain ornaments, these veils weigh down on me," by seeking out light fabrics and a sober attire.[7]

Lyotard's reduction of social to discursive practices dovetails with the earlier discussed thesis of the adequacy of language to intelligibility. This joint penchant, nurtured in the twentieth century within the divergent traditions of structuralism, poststructuralism, analytic philosophy, and hermeneutics, is opposed by the emphasis on practice and practical rooting trumpeted in pragmatism, Wittgenstein, and *Being and Time*. This front stresses (1) the integrity of doings as nonlinguistic, nonspeech-act, and nonsymbolic occurrences carried out by a socially regimented and molded expressive body, and (2) the inability of language to capture exhaustively the understanding through which these performances proceed. It thus repugns the hypostatizations of abstract orders, the overintellectualizations of action, and the excessively determinate specifications of intelligibility that characterize linguistifying approaches. Acute practical sensibility is also a central feature of the two most extensively developed theories of practice, to which we now turn.

The Governing of Action and the Organization of Practice

One of the two principal topics examined in this chapter is the unity of action and practice envisioned in certain analyses of practice. As noted, the consolidation of action and practice rests on the realization that practices are composed of the doings and sayings of individuals. This realization suggests that the organization of practice and the governing of its constituent actions must somehow coincide, intermesh, or be structured similarly. This idea of a coincidence or common structuring of the organization of practice and the governance of action in turn provides a red thread along which to compare and evaluate different accounts of practice. Of course, the supposition of commonality does not imply that all properties of practices are inscribed in the determination of their constituent actions. Features of practice that are not reflected or transcribed in this determination generally pertain to the interlocking of *different* integrative practices in social life. (The interlocking of dispersed practices is generally inscribed in understandings.) The present chapter will focus on individual practices, however, and mostly leave unconsidered the wider nexus of practices in which any integrative practice is embedded.

Bourdieu's Theory of Practice

Bourdieu's theory of practice and Giddens's theory of structuration boast profound visions underwriting unified positions on a wide variety of social theoretic issues. Whereas Bourdieu deftly captures practical difference and highlights the role of the body and the constitutive significance of intelligibility in social life, Giddens accents regularity and routine and the omnipresence of power in everyday interaction. Both, moreover, intend to let stand the practical nature of action, practice, and intelligibility. Both, however, fail to acknowledge this nature fully and uphold the integrity of social science only by clinging to the thorny road of intellectualism.

Bourdieu's central intuition is that practices (in his language: the activities, or "games," found within specific "fields," or "domains of practice") are self-perpetuating. Participants's actions are produced by dispositions (habitus) that, in being acquired under the objective conditions that characterize existence in the context of certain practices, generate actions that reproduce and perpetuate the practices and conditions. Practices, conditions, and dispositions are mutually calibrated. Left to their own devices, consequently, practices are self-propagating histories of activity. Indeed, given Bourdieu's devaluation of consciousness (see subsequent discussion), the main source of social change in his theory appears to be the collision of different practices and nexuses thereof.

The objective conditions under which dispositions are acquired come in two basic types. The first are statistical regularities, for example, price curves, chances of access to higher education, laws of the matrimonial market, and frequency of holidays. Bourdieu's idea is that within a given field, thus more generally in a social world characterized by such and such practices in this or that field, such parameters as prices, chances of access, matrimonial laws, and frequencies of holidays are statistically regular. The second type of objective condition is the "space" of social groups ("classes"). The groups involved are (1) defined by the possession of certain cultural, symbolic, and material capitals, thus by age, education, occupation, rank, sex, family, possessions, and the like, and (2) marked by the pursuit of certain actions and goods (life-styles). One extremely important type of group is economic classes, which are ensembles of individuals occupying similar positions within the distribution of economic capital. Bourdieu's idea again appears to be that within a specific field, and thus more generally in a social world characterized by such and such practices in this and that field, the particular prices, chances, laws, and frequencies individuals face reflect both the groups to which they belong, that is, their "position" in group space (thus the capitals they possess) and relations among these groups.[8]

In a world characterized by such and such practices in this and that field, which behavioral dispositions a person acquires depends on his position in group space and his location within statistical regularities.

These dispositions, in turn, generate actions that perpetuate the practices and conditions involved. In *Outline of a Theory of Practice* and *The Logic of Practices*, Bourdieu accounts for this harmony by arguing that dispositions are structured homologously with the conditions under which they are acquired (see note 16 for his tack in other texts.) This thesis captures an important truth, namely, that people's inclinations and penchants reflect the social circumstances in which they live. That, for instance, American inner-city blacks tend to treat whites irreverently and hostilely obviously reflects the circumstances of poverty and antiwhite culture in which their inclinations are molded. However, the claim that people's dispositions systematically produce actions that perpetuate the practices and conditions within which they are acquired is a stronger proposition, one that rests on Bourdieu's expansive construal of dispositions and his analysis of them as structured homologously with these conditions.

The dispositions that constitute habitus are responsible not only for behaviors, but for thoughts, motivations, and perceptions as well. Although Bourdieu never comes out and says this, habitus is responsible for more or less everything that goes under the label "mind/action." One finds him characterizing habitus, accordingly, as "mental dispositions," "schemes of perception and thought," and "cognitive and motivating structures." These expressions, in conjunction with his references to habitus as "mind" (in *OT*), also suggest that he intends the notion of habitus to replace that of mind in the analysis of human life. The dispositions that constitute habitus, moreover, are "bodily schemes." Bourdieu labels them so in part to emphasize that the operations of habitus are carried out by "bodily gymnastics" and also transpire both nonconsciously and automatically. Mental dispositions and the like, as a result, are "inscribed" in the body.

Since habitus reflects the objective conditions under which it is incorporated, what a person is and does are ultimately beholden to such conditions. Habitus, however, does not simply mechanically regenerate the actions encountered in the contexts of inculcation and, in this way, perpetuate practices and conditions. The actions habitus produces are also sensitive to the ever changing situations of action. Appropriate in their specific circumstances, yet continuous with the practices in which the bodily schemes were acquired, these actions perpetuate practices by extending them to new circumstances. Objective conditions determine action, consequently, by laying down limits to the improvisations of habitus:

Through the habitus, the structure which has produced it [the habitus] governs practice, not by the process of mechanical determinism, but through the mediations of the orientations and limits it assigns to the habitus' operations of inventions. (*OT*, p. 95)

Habitus plays two basic roles vis-à-vis action: generation and selection. The bodily schemes generate actions, that is, causally produce them. (Bourdieu repeatedly refers to actions and artifacts as "products" of habitus.) The bodily schemes also select the actions to be generated. The first of these two functions is highlighted when Bourdieu refers to habitus as "dispositions," the second when he describes it as the "principle" of practice. The applicability to human behavior of the dual dimensions of causality and intelligibility has been the subject of intensively vigorous disputation over the past century and a half. I have discussed Bourdieu's attempt to assign them to one and the same "mechanism-structure" elsewhere, and will not repeat that discussion here.[9] Instead, I will concentrate on habitus as the principle of practice.

To conceptualize habitus as the principle of practice is to credit it with selecting the actions people perform. According to Bourdieu, the bodily schemes select actions that are "sensible" and "reasonable" (e.g., LP, pp. 14, 50). "Sensible and reasonable" means, first, that an action is appropriate given the situation and the functions action is to fulfill in that situation. It means, second, that the action makes sense to the actor, that is, to someone whose schemes of action, perception, and thought have been formed within certain practices and conditions (e.g., OT, p. 79; LP, p. 66). The actions that habitus selects thus make sense given the situation and also given the objective conditions and practices familiar to and inhabited by the actor.

This means that the actions the bodily schemes select will also be sensible and reasonable to other actors who have matured within and become accustomed to the same practices and conditions. Indeed, since, as we will see, habitus and world share a common structure, habitus's schemes of understanding and perception effect a

common-sense world, whose immediate self-evidence is accompanied by the objectivity provided by consensus on the meaning of practices and the world, in other words, the harmonization of the agents' experiences and the constant reinforcement each of them receives from expression . . . of similar or identical experiences. (LP, p. 58)

The homology of the habitus of actors who grow up and live amidst the same practice-established objective conditions also ensures that the actions they individually perform add up to regular, unified, and systematic (LP, p. 59) social practices. Since these practices are structured homologously with actors' dispositions, which in turn are structured homologously with the objective conditions under which they were inculcated, the practices perpetuate the objective conditions involved. So, for example, in participating in the same practices and facing the same objective conditions, members of a given economic class acquire a "class" habitus that ensures (1) that their actions are sensible and reasonable to one another, (2) that they know their way about the practices characteristic of this class, (3) that their individual actions add up to coherent "class" practices, and (4) that these practices perpetuate the practices and objective conditions under which class habitus was acquired.

Most important, action intelligibility, the selection of action, conforms to what Bourdieu calls "practical logic." Practical logic is a representation of the principles the bodily schemes observe in selecting actions. In any context, according to Bourdieu, habitus selects actions by producing (1) a definition of the situation of action, which assigns meanings to objects, persons, and events, delineates a probable upcoming future, and prescribes "objective potentialities" (things to say and not to say, to do and not to do); and (2) a definition of the functions of action in that situation (OT, p. 142; LP, p. 267). (Habitus then generates the action that best fulfills these functions.) Practical logic describes the principles the bodily schemes follow in rendering these definitions.

Bourdieu's account of these principles ultimately rests on the Saussurian thesis that meaning devolves from systems of differences. Habitus, consequently, composes definitions of these two sorts via systems of differences, which in Bourdieu are oppositions. Examples are day/night, high/low, wet/dry, going forward/going backward, and hot/cold. The oppositions habitus wields form families, each based on a fundamental opposition. In *Outline of a Theory of Practice*, the fundamental oppositions are identified as opposed movements of the body such as going to the left and going to the right, going in and coming out, sitting and standing, and going up and coming down (OT, p. 119). In *The Logic of Practice*, Bourdieu appears to replace the plurality of fundamental oppositions based on bodily movements with the single opposition between male and female (e.g., LP, pp. 78, 146, 223).[10] In any event, families of oppositions are built up by the association of oppositions with the fundamental ones, either directly or via other oppositions that have already been associated with them. Wet/dry, for instance, might be directly associated with female/male, while lower/upper parts of the house might be associated with female/male by way of association with wet/dry, which is already associated with female/male. For these oppositions to be associated means semantically that as female is to male, so too is wet to dry and lower part of the house to upper part of the house.

Practical logic, then, is a description of the principles that govern the assignment of meaning to the situations and functions of action through the construction and application of families of opposition. The ways families of opposition are built and applied are fundamentally analogical, play upon polysemy, work via "uncertain" and "false" abstraction based on overall resemblance between entities, and represent a great loss of rigor in comparison to "logical" logic. Although Bourdieu never spells this out in detail, it is reasonable to suppose that objects, persons, and events are assigned meanings by being subsumed into the families of opposition. For instance, the upper part of the house might be associated with one side (dry) of an opposition (wet/dry), become opposed to something else (lower part of the house) through analogy with that opposition (lower part/upper part::wet/dry), and thereby come into association, along with its opposite, with the entire family of analogously opposed "realities" (*LP*, p. 88) of which the opposition to which it has been attached is part (wet/dry::female/male . . .). The meaning the upper part of the house then acquires lies in the difference from its opposite, which sounds out of the combined significance of the realities with which it has been associated (dry, male) in their analogous differences from the realities to which they are opposed (wet, female). It seems further reasonable to presume that probable futures as well as prescribed and proscribed actions are delineated on the basis of assigned meanings, and that the functions of action follow from these assignments and delineations. In this way, the selection of action is structured by oppositions.

I note that often what it is for an action to "best" fit the definition of the functions of action is for it to be the action that maximizes the actor's capital. Bourdieu argues that practices generally follow the logic of capital maximization, not only material capital, but social and symbolic as well (*LP*, p. 16; *OT*, p. 183). With regard to the governing of action, this presumably means that habitus observes this logic in selecting which of those actions that could fulfill the functions of action is the one performed.

It is important to emphasize that habitus's work of selecting actions transpires nonconsciously; an actor is not aware of the definitions, oppositions, and families involved. Bourdieu concedes, in fact, that talk of selection and definition is overly intellectualist. It implies that behavior is governed by mechanisms that operate nonmechanistically over representations, when in fact no such mechanisms are at work. The principles and operations spelled out in practical logic are "performed directly in bodily gymnastics" (*LP*, p. 89; *OT*, p. 117), without mediation of representations or cognitive operations. They are immediately and nonrepresentationally acted out by the bodily "practical transfer of incorporated, quasi-postural schemes" (*LP*, p. 92). Even the "logical transformations" attributed to practical logic – for instance, the joining and separation of oppositions, and the transferal of families of oppositions among different realities – take the form of bodily movements, such as tying and cutting, turning right or left, and turning around (*LP*, p. 92).

To turn now to the organization of practices, the families of oppositions not only structure the selection of actions but also organize social practices. Bourdieu contends that practices in a variety of fields, for example, cooking, agriculture, gift giving, marriage, and the women's day, possess a temporal organization structured by the same families of oppositions that structure habitus (*OT*, pp. 130–157; *LP*, book II, chap. 3). The Kabylian cooking cycle, for instance, exhibits a temporal organization of what is cooked when that is structured by such oppositions as boiled/roasted, sweet/salt, bland/spiced, and wet/dry. Bourdieu further and famously shows that the layout of the built environment similarly exhibits an opposition-based organization. This is true of small-scale settings such as the house and its spaces (see *LP*, pp. 271–283) as well as the wider arrangements of towns and agricultural fields. It turns out, therefore, that the same families of oppositions homologously structure habitus, practices, and the layout of the world.

Two questions arise concerning the scope of oppositional structure. The first is whether all practices, or all dimensions of practices, are so structured. I'm not sure what the answer to this question is for Bourdieu. He occasionally remarks that the oppositional structure discussed in the two texts under consideration applies solely to ritual (e.g., *LP*, p. 18).[11] Although this might suggest that oppositional structure characterizes traditional societies alone, Bourdieu can also be read as arguing that practices in all societies exhibit a significant dimension of ritual. The systems of opposition discussed in *Outline* and *Logic* rest on certain fundamental differences found in all societies: either opposed movements of the body or the ultimately biological distinction between male and female. If, consequently, oppositional structure based on these differences signals ritual, then the ubiquity of the fundamental oppositions entails the omnipresence of ritual. As noted in an earlier footnote, moreover, Bourdieu also maintains that practices and objective conditions in class-divided societies are structured by oppositions, all of which refer back to fundamental ones (the "mythic roots").[12] Henceforth, accordingly, I will assume that oppositionality characterizes the full gamut of social practices. This interpretation is indirectly confirmed by the oppositional structures that Bourdieu unearths and works with in his work on contemporary France, the academy, language, and power.

Second, I indicated earlier that Bourdieu's theoretical texts contend that habitus is structured homologously with objective conditions that are established with the reign of particular practices. As I have now explained, however, habitus is in fact homologous with the practices, not the objective conditions, concerned. Bourdieu systematically misrepresents his position whenever he writes of a homology between objective conditions and bodily schemes. Indeed, it is not at all obvious how statistical regularities and group space can in principle be structured by the systems of oppositions (going in/going out, dry/wet, hot/cold, up/down, etc.) that structure habitus. Suppose that certain statistical regularities having to do with prices, frequencies, and the like track the distribution of economic (and/or other) capital. The "structure" of these regularities would then consist of conjunctions of the values assumed by different socioeconomic variables, together with whatever mathematically or linguistically describable relations characterize either the regularities or sets of conjuncts. These conjunctions and relations are clearly not homologous with the opposition-based structure of habitus's definitions of the situation. Bourdieu is right, of course, that people who belong to a given economic class tend to face similar frequencies of holidays and the like. He may also be right that such people tend to act in similar ways, at least insofar as their actions are keyed to these conditions. But these facts in no way suggest that dispositions, on the one hand, and collocations of either statistical regularities or specific values of socioeconomic variables, on the other, are structured alike. It is logically wiser to postulate that the dispositions are structured homologously with the manifolds of action they produce.

And, in fact, Bourdieu makes no attempt, in *Outline* and *Logic*, to demonstrate that these objective conditions are so structured. He shows, instead, that practices are structured oppositionally. Bourdieu's programmatic assertions of a homology between habitus and objective conditions are confined to passages that sketch his theory abstractly. Whenever (in these works) he spells out the specific structure of habitus, the homology he establishes holds between bodily schemes and practices (see note 16 on other texts). Habitus, consequently, is not acquired "within," and does not directly perpetuate, objective conditions. Rather, it is acquired through participation in the social practices that it perpetuates through the actions it generates; and in perpetuating these practices it thereby also perpetuates the objective conditions established through the existence of those practices.

Finally, it might be objected that this rendering of Bourdieu's account of practices neglects a key dimension of their structure, namely, the various forms of capital. This is true only nominally. The existence and pursuit of capital depend on actors' interpretations of people and entities. This holds not only of nonmaterial capitals, as Bourdieu acknowledges, but also of the material sort, since land, cattle, metal coins, and the like amount to capital only given appropriate interpretations of them. Because, therefore, habitus underlies actors' interpretations of the world, the existence of all capitals ultimately rests on the oppositions structuring these dispositions. These oppositions, moreover, in structuring the selection of action, likewise structure the intelligibility of the strategies people deploy in pursuing capital. Only the principle, Maximize Capital!, swings free of oppositional structure – even though of course, actors' understanding of the principle does not.[13] The capital dimension of practice, consequently, is mostly structured by the families of opposition. So the preceding analysis overlooks capital only in not discussing it explicitly. I do not mean to suggest that capital is not a key component of Bourdieu's account, only that it, like just about everything else in his theoretical account of practice, is in the end a reflection of habitus.

Giddens's Theory of Structuration

Giddens's central intuition, much like Bourdieu's, concerns the perpetuation, or extension, of practices over time-space. Space-time extension is made possible through the mediation of structures, which are at once the conditions of practices and something reproduced by them. Practices, moreover, form interlocking nexuses called "systems." The structural properties of systems, like those of practices, are at once conditions of the systems' constituent practices and something renewed by those practices.

Practices, furthermore, are composed of individuals' activities. Social reality is a tangle of streams of activity that compose practices with structures that are both the condition and outcome of those practices. Structures, consequently, must also somehow be the condition and outcome of individual activity. Giddens proposes, accordingly, that individuals "draw upon" the structures of practices, thereby renewing the structures and participating in and perpetuating the practices. In so doing, actors also contribute to the perpetuation of the social systems composed of these practices: The structures "drawn on in the production and reproduction of social action are at the same time the means of system reproduction" (CS, p. 19). Actions, practices, systems, and structures thus form tightly bound complexes.

Social structures, meanwhile, are composed of rules and resources. Giddens recognizes two sorts of rule: codes, which determine the meanings of things, and norms, which determine right and wrong (legitimation). All practices, Giddens argues, are governed by rules of both sorts. More abstractly, rules are methodically applied generalizable procedures of action implicated in the practical activities of daily life (CS, p. 21). Unlike in the previous chapter as well as in

Bourdieu's account, rules are not explicit formulations. They are "procedures of action," which Giddens likens to formulas and typifications (*CS*, pp. 20–22, 29). Such procedures, he writes, cannot be exhaustively analyzed in terms of (formulated) contents (e.g., "Do X"), because *"rules and practices only exist in conjunction with one another"* (*CP*, p. 65, italics in original). Rules are methodically applied, moreover, in being repeatedly observed. Repeated observance is, in turn, responsible for action regularities in daily life. Rules, consequently, are "recursively implicated" in regularities.

Giddens's notion of a resource presents greater interpretive difficulties. What he has in mind with the notion seems clear: Resources are the medium through which social power is exercised, where social power is the capacity to bring about changes when doing so depends on the actions of others (*CP*, p. 93). Functionally, therefore, resources are that, through the utilization of or reliance upon which actors can bring others to perform actions through which specific outcomes are secured. Giddens defines resources, however, as capabilities that generate commands either over persons or over objects and other material phenomena (*CP*, p. 100; *CS*, p. 33). Examples of resources that generate commands over persons (authoritative resources) are the organization of activities, the coordination of actors, and aptitudes and capabilities (*CS*, pp. 258, 260–261). Examples of resources that generate commands over things (allocative resources) are wealth, technologies, raw materials, and land (*CP*, p. 104; *CS*, p. 258). The first problem with this definition is that it is not compatible with the functional specification of resources. The medium through which a capacity is exercised cannot itself be a capability; at the least, it must be the exercise of that capability. A second problem is that, if resources, as the definition has it, are capabilities, the entities Giddens cites as examples of resources (apart from "aptitudes and capabilities") do not obviously count as such. Indeed, his examples of allocative resources more closely resemble the objects and material phenomena, commands over which are said to be generated by resources.

I am not sure how to expurgate these inconsistencies. One could hold fast the functional specification of resources. In this case, however, a person exercises power not, strictly speaking, through phenomena of the sorts Giddens cites, but through his or her command (i.e., control or possession) of them. What a person draws on in seeking outcomes that depend on others' actions, that through which the capability to achieve these outcomes is realized, is her command over persons and things. On this first option, consequently, commands, and not the cited phenomena, count as resources. On the other hand, one could stand by the specification of resources as capabilities. In this case, however, Giddens's intentions would seem best achieved by analyzing resources as capabilities that arise from – and not: that generate, as his definition has it – commands over persons and material phenomena. Interestingly enough, Giddens appears to characterize resources so in the introduction to *The Constitution of Society* (p. xxxi).

In any case, structures are sets of rules and resources, which are at once the medium in which practices are carried out and the renewed result of their execution. Since practices compose systems, the structural properties of social systems are likewise sets of rules and resources (together, in this case, with relations among these sets) that are both medium and result of system practices. What's more, since practices and systems are composed of actions, the ultimate reason why rules and resources structure practices and systems is that actors draw on rules and resources in their interactions. In doing so, they perpetuate the practices of whose structure the rules and resources are elements, and thereby also help reproduce the social system composed by these (and other) practices.

Giddens's account of how actors "draw on" rules and resources reveals an ontological bifurcation. On the one hand, rules and resources, along with the sets they form and the relations among these sets, live a "virtual" existence. This means that, like the system of differences composing signifieds in Saussurian linguistics, they exist outside space and time. Giddens draws back, however, from Saussure's hypostatization of the virtual realm. He manages this by also embedding structure in actors' practical consciousnesses. The tacit knowledge that orients people in their interactions is a knowledge of rules (*CP*, p. 68; *CS*, p. 21). A parallel claim should, architectonically, apply to resources, although Giddens never explains how this would work (I will offer an explanation of sorts in the next section). In any event, actors "draw on" rules and resources in the sense that social interactions are conducted on the basis of a practical understanding that enmeshes rules and resources. Thus, in addition to their virtual existence, structures also have a space-time being, consisting in "memory traces" of how things are to be done (*CS*, p. 25). Giddens writes, accordingly, that "[s]tructure has no existence independent of the knowledge that agents have about what they do in their day to day lives" (*CS*, p. 26).

Qua virtual order, sets of rules and resources are recursively implicated in the regularities of the spatiotemporal manifold of social interaction. Social scientists reconstruct structure, as a result, on the basis of these regularities. *Qua* spatiotemporal order, on the other hand, structures are present in actors' knowing how to go on. But structure *qua* practical consciousness cannot for the most part be directly accessed. Whereas discursive consciousness is everything actors understand, grasp, or know and can express in words, practical consciousness is everything they understand and know but cannot formulate verbally. There is no absolute line between the two since discursive accessibility can vary with education, upbringing, and the like (*CS*, p. 7). At least most of the rules drawn on in social interaction, however, are fixed in practical consciousness. (Giddens does not say one thing or another about resources.) Structure is thus primarily accessible only as a reconstructible virtual order implicated in the space-time manifold of practices, and not as a recollectable order present in these practices.

Rules and resources are not the only "determinants" of action. What people do also depends on their reasons and wants. By "reasons," Giddens means the grounds on which people unspeakingly and continuously understand their activity to rest (*CS*, pp. 5–6, 376). Reasons are akin to the teleoaffective orders discussed in the previous chapter, differing because actors cannot easily access all components of such orders. By "wants," on the other hand, Giddens signals motivations rooted in the unconscious. Like practical consciousness, the unconscious is inaccessible to explicit awareness, in this case because "a negative 'bar' of some kind inhibit[s] its unmediated incorporation within the reflexive monitoring of conduct" (*CS*, p. 49). The unconscious houses deep-seated "general wants" for survival, the avoidance of anxiety, and the preservation of self-esteem (*CS*, pp. 177, 57). Since routine standardly provides for them, these wants often underlie a generalized motivational commitment to the integrity of routine practices (*CS*, p. 64; *CP*, p. 124). Moreover, since most of daily life is routine, general wants are usually satisfied and actions not directly motivated by them. Only in critical situations, when routine is disrupted, do general wants directly give rise to behavior, which seeks to restore the "ontological security" maintained in routine.

Whereas reasons are lodged in discursive consciousness, and wants are rooted in the unconscious, the rules and resources that compose structure are housed in practical consciousness. Like unconscious wants, they are inaccessible to discursive formulation. Unlike wants and reasons as well, they are constantly determinative of human activity. They alone, moreover, form the omnipresent medium and outcome of actions and practices. Reasons and wants swing free of structure.

Unlike Bourdieu, Giddens does not offer a complete account of action. Drawing on the work of Erving Goffman, he writes that actors negotiate social interactions on the basis of "frames." Frames are clusters of codes and norms with which people (1) make sense of events, states of affair, and what others do and (2) understand themselves and others both to possess certain rights and obligations and to be subject to specific sanctions if they perform certain actions (*CS*, p. 88). When actors share frames, interactions are mutually intelligible and coherent. Clusters of rules, however, do not specify more precisely what makes sense to interacters to do. A plurality of actions is compatible with at least most rule-permeated understandings of circumstances, others' activities, and participants's rights and obligations. Wants cannot play the pendant of Bourdieu's maximization of capital and Heidegger's signifying, for they do not engage with action in routine situations. Reasons alone would be capable of "selecting" which circumstantially and normatively appropriate actions are performed. Giddens maintains, however, that actors do not need reasons for routine activity (*CP*, pp. 218–219). Reasons, consequently, do not play a comprehensive role in his account. Perhaps he believes that the rules governing routine circumstances do in fact specify behavior.

In any case, in conducting their interactions on the basis of clusters of rules (and resources?), actors renew rules and thereby participate in and perpetuate the practices of whose structures the rules are elements. So long as people observe the rules, whatever they do maintains the practices concerned, regardless of whatever reasons and wants they may have for their actions.

Finally, the differentiation of two sorts of rule and two types of resource underwrites a fourfold division in institutional orders: symbolic, political, economic, and legal. These orders are not distinct sets of practice. The political order is not composed of practices a, b, and c, the economic of x, y, and z, and the legal of l, m, and n. All practices have semantic, political, economic, and legal dimensions. The semantic dimension is the order of codes that structure a practice and are drawn on by actors in the interactions composing it. The political and economic dimensions, meanwhile, are the structuring and drawn-on orders of authoritative and allocative resources respectively, while the legal dimension is the order of norms. Each dimension, taken by itself, is an abstraction, for in ongoing life rules and resources of different sorts operate in conjunction and are drawn on by actors as interdependent sets. Institutional orders are thus abstract dimensions through which practices course, not distinct, differentiated regions of social systems.

Practical Understanding and Practices

Practice as spatiotemporal manifold and practice as do-ing are two aspects of one and the same reality of human praxis. Activity and performance are unified by a single order, consequently, that governs actions at the same time that it organizes practices. In Bourdieu, this order is a system of oppositions, which structures both a practice's space-time organization and the selection of actions by its participants' habitus. In Giddens, the rules and resources actors draw on when interacting within a practice also are the medium through which the practice extends itself over time and space. In my account, finally, the common order is composed of understandings, rules, and teleoaffectivities. The current subsection argues for the superiority of the latter conception.

In terms of the triad understanding/rules/teleoaffectivity, Bourdieu collapses the organization of practices entirely into understanding. Rules – from proverbs and sayings to coherent corpora – are governed by the oppositions that structure habitus. Some rules are attempted formulations of the intelligibility already underlying practices: In "making explicit and objectifying . . . the generative schemes in a grammar of practices" (OT, p. 20), they reinforce and consolidate the determination of action by the bodily schemes. Other rules do not make explicit already operative "principles," but instead introduce imperatives by reference to which activity should henceforth proceed. Since habitus, however, governs the formulation of such rules, their intelligibility is structured by the oppositions. Rules, consequently, are not an independent determinant of practice.[14]

Bourdieu, moreover, dismisses a teleoaffectivity independent of understanding as irrelevant to both the determination of action and the organization of practices. Habitus generates not only behavior, but thoughts, perceptions, and motivations as well. Thoughts, for instance, are mere accompaniments of practice and not the reasons, motives, or causes of it (OT, p. 21). That it phenomenologically appears to people that they orchestrate their behavior is an illusion: "It is, of course, never ruled out that the responses of the habitus may be accompanied by a strategic calculation tending to carry out quasi-consciously the operation the habitus carries on in a quite different way" (LP, p. 53; OT, p. 76). Actions and practices are also not determined by ends and projects:

Even when they appear as the realization of the explicit . . . purposes of a project or plan, the practices produced by the habitus, as the strategy-generating principle enabling agents to cope with unforeseen and ever-changing situations, are only apparently determined by the future. (OT, p. 72; cf. LP, p. 61)

Indeed, projects and plans are constructions of social analysts (LP, p. 81).

As these quotations also suggest, Bourdieu operates with a two-level conception of "mind": states of consciousness and nonconscious bodily schemes. I have emphasized, however, that cognitive and emotional conditions (thus plans and projects) are not states of consciousness. Teleoaffective orders can structure a person's ongoing behavior even when this behavior is a stream of reactions. (Incidentally, Bourdieu's theory in effect construes *all* human behavior as reactive; see the references to spontaneity at *LP*, p. 56, 104, and *OT*, p. 80.) These orders might indeed be graspable only as post hoc "reconstructions." But this does not transform them into inventions of social analysts. These reconstructions articulate an order that is present at the time of do-ing and that bears upon do-ing by structuring what is signified as the thing to do. Consequently, Bourdieu's denial that an independent teleoaffectivity helps structure action and practice rests on an inadequate conception of this aspect of life.

The collapse of practice organization into practical understanding generates its own problems. Bourdieu implies that his account of practical logic is a theory of practical understanding in the sense Wittgenstein speaks of understanding (see *LP*, p. 18), that is, knowing how to go on. This is reflected in his frequent descriptions of habitus as "practical sense," as "the sense for the game," and as a consortium of different "senses," for example, those of beauty, duty, and reality, and business sense, common sense, and moral sense. Somehow, consequently, Bourdieu's practical logic details general principles of the know-how whose exercise is the operations of motor schemes and automatisms. As discussed in Chapter 2, however, practical understanding cannot be adequately expressed in words. This conclusion problematizes any "theory" that seeks to make explicit its contents or principles. An adequate practical logic, consequently, is an impossibility. It is impossible to construct an adequate representation of the "selection" of intelligible actions, of the principles that determine how to proceed intelligibly, to the extent that action arises from actors' knowing how to go on.[15]

Curiously, Bourdieu appears to agree. In harmony with a Wittgensteinian sensibility, but most provocatively given his own procedure, Bourdieu repeatedly insists that any theory of understanding is an irredeemable fiction! Its models are mere "theoretical equivalents" (*OT*, p. 11) of "the real mechanics of the schemes immanent in practice" (*OT*, p. 20). Not only are these "equivalents" not the real mechanisms; they are essentially falsifications that destroy practical principles as such (*LP*, p. 11). Indeed, the very idea of a practical logic is a "contradiction in terms" (*LP*, p. 92). Bourdieu maintains this not so much for the reasons articulated in Wittgenstein's writings, *viz*, the manifest inability to formulate practical understanding adequately in words. He more often writes that it is the bodily nature of practical understanding that vitiates any attempt to chart practice's principles in propositions and representations:

> But the language of overall resemblance and uncertain abstraction is still too intellectualist to be able to express a logic that is performed directly in bodily gymnastics. . . . Rites, even more than most practices, might almost be designed to demonstrate the fallacy of seeking to contain in concepts a logic that is made to do without concepts; of treating practical manipulations and bodily movements as logical operations; of speaking of analogies and homologies . . . when it is simply a matter of practical transfers of incorporated, quasi-postural schemes. (*LP*, pp. 89, 92)

An account of practical logic does not explicate the real understanding that underlies practices, but instead describes practices "as if" they were governed in certain ways (*OT*, p. 77). This means that Bourdieu's analysis of practical logic – via systems of oppositions and such principles as uncertain abstraction, the logic of overall resemblance, and the maximization of capital – no more yields the true principles of practice than *any other account* does (*LP*, p. 11).

One naturally wonders, then, why Bourdieu expends so much effort constructing a meticulously rich and marvelously suggestive theory – when the details are inherently unreal and the implications likely deceptive. His reasons lie in the scientific drive to explain. I believe Bourdieu regards abandoning the effort to analyze practical understanding as tantamount to renouncing the scientific call to account for practices. As if in response to his own eloquently expressed skepticism about the possibility of a theory of understanding, he rejoins that the only way to "give an account" of practices is to construct such a theory (*LP*, p. 92; cf. p. 89). His own theory is an attempt to give a coherent and economic account of the observed facts (*LP*, p. 11). Despite its falsifying intellectualism, the cause of science requires him to press on and construct it.

If the goal is to "give an account" of practices, it is not clear why a theory of practical understanding must be a component. According to the analysis of the previous chapter, practices are organized by understandings, rules, and teleoaffective structures. This analysis, regardless of its adequacy, counts as an "account" of practices, to wit a "scientific" (*wissenschaftlich*) one. Bourdieu's doggedness that an account of practices contain a practical logic reflects the dismissal of rules and independent teleoaffective orders from his analysis. Understanding alone, on his view, organizes practices. If, consequently, understanding were unanalyzable, the resulting account of practice, in particular the homology between habitus and practices, would lack content; it would stand merely as a tantalizing, but ultimately empty, picture of the nature of human activity. In order to give the picture more substance, as a result, Bourdieu is forced against his own convictions and admonitions to offer an account of the unaccountable.[16] A better solution would be to readmit rules and teleoaffective orders. This move is blocked within Bourdieu's scheme, however, by the omnipotence of habitus.

Questions also arise about Bourdieu's contention that the selection of sensible and reasonable ways of proceeding operates via oppositions. As an account of action intelligibility, practical logic competes with Heidegger's depiction of the teleoaffective structure of signifying. Bourdieu implies that one reason for favoring his analysis of habitus is that it offers an alternative to the reigning either/or of mechanical causes or conscious conditions. "The *habitus* is a spontaneity without consciousness or will, opposed as much to the mechanical necessity of things without history in mechanistic theories as it is to the reflexive freedom of subjects 'without inertia' in rationalist theories" (*LP*, p. 56). Heidegger's signifying, however, equally represents an alternative to cause-or-consciousness. So neither account is preferable on this ground alone.

Crucially differentiating the two accounts is the fact that intelligibility in Bourdieu is articulated by systems of opposition, whereas in the previous chapter it was portrayed as articulated by ends, projects, tasks, states of affairs, and how things matter. An implication of this difference is that on Bourdieu's account actors articulate intelligibility with concepts distinct from those of the theory, while on my analysis the two sets of concepts overlap and are otherwise exchangeable. In Bourdieu's theoretical texts, for instance, oppositions do sometimes appear in sayings (e.g., *LP*, p. 90, middle of the page), but actors do not understand and explain their practices by invoking oppositions and associations thereof. (In other works, Bourdieu sometimes analyzes opposed ordinary language terms and aligns them with fundamental oppositions.) Bourdieu's account thus hypothesizes an underlying stratum of intelligibility, in the sense of an intelligibility whose nature and contents differ from those actors live with and through in ongoing existence.

This difference is a positive reason to favor the Heideggarian account. Perhaps it is an outmoded prejudice, but I remain convinced that people's common explanations of self and others on the whole correctly articulate why they act as they do. I do not imply that people never err, or that getting straight about behavior does not sometimes require overcoming considerable barriers. I mean simply that people do not *systematically* offer inadequate explanations due to a fundamental divergence between their schemes and the real principles of action. Since signifying, unlike the structure of habitus, is accessed by actors and articulated in their accounts of themselves, Heidegger's analysis alone respects this principle.

In fact, some of Bourdieu's formulations of the task of a theory of practice suggest that he, too, should not countenance a fundamental divergence of this sort. The theorist, he writes, should "construct their [practices'] generative principle by situating [himself] within the very movement of their accomplishment" (*OT*, p. 3) – "by situat[ing] [himself] *within* 'real activity as such,' that is, in the practical relation to the world, the preoccupied, active presence in the world through which the world imposes its presence" (*LP*, p. 52; cf. *OT*, p. 96). As Bourdieu's occasional phenomenologically adroit descriptions demonstrate, the locus of the "practical relation" to the world, where the world imposes itself and the preoccupied agent consummates her dealings with it, is lived experience. Accordingly, the structure of action intelligibility revealed by the procedure of "situating oneself" is a structure of lived experience, accessible to actors and most likely articulated in their daily accounts of self and others. "Situating" does not disclose an intelligibility that is systematically structured at variance with actors' self-explanations. The same conclusion is reached even more directly once it is realized that taking seriously the desideratum of situating oneself in "the practical relation to the world" implies taking seriously the explanatory accounts people offer of themselves *in* their ongoing practical dealings with one another. Hence, becoming a "theoretical subject" (*LP*, p. 145) of practices implies treating the conceptual scheme actors wield in their self-explanations as adequate in principle to the articulation of intelligibility. Bourdieu, therefore, is poorly served by this description of the theoretical task. His enterprise is not one of theoretically reconstructing the practical from within it.

It is instead one of theoretically reconstructing the practical on the basis of its products. It is only from the "outside" that action intelligibility can be attributed an underlying structure divergent from people's everyday accounts. The truth of this claim is indirectly confirmed by the one real argument Bourdieu offers for the existence of an underlying, opposition-based intelligibility. Postulating such a structure, he claims (e.g., *LP*, p. 95), is the only way of accounting for the homologous oppositional structure of practices in different social arenas, such as agriculture practices, cooking practices, women's daily activities, and the organization of the day. In other words: That which selects and generates actions that compose different but homologously structured practices must be structured the same. That this structure diverges from actors' explanations of these same actions is neither here nor there.

Bourdieu's assumption that the structure of a product must be homologous to that of the "mechanism" producing it can be challenged. For instance, one can rebut the claim that the intelligibility governing the use of closings in letters is structured by the oppositions with which post hoc investigation can catalog the addressees of different closings (see *LP*, p. 293 n. 4). I will consider, however, a different difficulty with Bourdieu's position, one related to the very idea of a theory of practice. By Bourdieu's admission, the diagrams he constructs of the yearly course or daily distribution of practices (e.g., *LP*, pp. 201, 220; *OT*, pp. 134, 150) falsify the practical organizations of the practices involved. The totalizing juxtaposition of oppositions only sequentially applied by habitus and never mobilized together as an ensemble, neutralizes the practical, situational attunement of their use and creates a host of relations such as simultaneity, symmetry, and, it should be added, homology that do not exist in practice (*LP*, pp. 83–84). This does not invalidate these diagrams as objects of scientific construction. But it does mean that conclusions cannot be automatically drawn, on the basis of their properties, about the structure of the "mechanisms" responsible for the practices they depict.

In particular, since the homologous organizations the diagrams reveal are products of the temporal totalization effected in these representations, it is not legitimate to read a homologous structure back into the governing mechanisms. The diagrams are in fact very misleading in this regard. For instance, choosing to diagram the day's activities and the courses of year-long practices with the same, proportionally divided sine wave is a theoretical decision that induces homologies between the structures of the practices and of the day's activities that are not in any obvious sense there in the sequences themselves. Even diagraming the yearly courses of, for example, agricultural and cooking practices with one and the same proportionally divided sine wave introduces homologies whose artificiality is based on the use of objective units of time not employed in the practices themselves. I am not claiming that different practices might not in some sense have homologous structures (or, again, that these diagrams are not of extreme interest). I maintain simply that Bourdieu cannot legitimately infer that habitus has an oppositional-based structure from the oppositional homologies exhibited in his diagrams.

Giddens replicates Bourdieu's folding of the organization of practices into practical understanding. Although he acknowledges teleologies and affectivities in the guise of reasons and wants, reasons and wants are not elements of structure; they do not contribute to the repetition and regularities of practices over time-space. Practices are structured solely by the rules and resources that are embedded in practical consciousness.

Earlier I discussed a difficulty with the definition of resources. Another problem besetting them is that drawing on resources is not distinct from drawing on rules. Resources, recall, are either commands over persons or things or capabilities that arise from such commands. On either construal, drawing on a resource involves exercising a command. Whence, however, commands over objects or persons? Charisma, emotional ties, the threat of physical violence, and the possession of special knowledge exemplify sources of commands without structuring power on Giddens's account. A source with such power is a person's location in the hierarchy of social positions, where a position is a social identity to which rights and obligations accrue (*CP*, p. 117). When social position is the source of command, an actor's drawing on a resource to induce others to perform certain actions involves relying on the codes and norms that define social identities and the rights as well as obligations attending these. Suppose, for instance, that a boss, drawing on the authoritative resource to orchestrate employee's actions that is based on her identity and rights as boss, instructs an employee to post a letter by courier service and the employee complies. Her capacity to determine the employee's actions rests on the codes and norms structuring business practices. This means, however, that her drawing on this authoritative resource really comes down to her drawing on the rules that structure the field of business interaction. And, in fact, this same conclusion holds for that wide variety of cases in which drawing on a resource involves exercising a command over things or persons that arises from social position. A parallel conclusion also holds, however, for any resource, the use of which embraces exercising a command that arises from emotional ties, charisma, knowledge, or force. For one person's exercising control over others by virtue of these phenomena rests on these other people's application of codes and norms to her. That, for instance, she is charismatic or threatening presupposes particular codes. In all cases, therefore, to draw on a resource is at bottom to draw on some set of rules in a specific situation. Possession of land, coins, raw materials, and the instruments and products of production, just like control over the organization of activities, the placement of individuals within them, and knowledge and aptitudes, rests on the codes and norms governing the practices in which these items are possessed and controlled.

Resources, consequently, do not merely structure practices in conjunction with rules (*CP*, p. 104). Their structuring effect is *nothing but* that of a conjunction of rules in the presence of or in relation to certain persons or material objects.[17] This implies, in turn, that the economic and political institutional orders are not distinct from the semantic and legal ones. The economic and political orders do not merely exist only in conjunction with the other two. They are nothing but particular orderings of the semantic and normative – particular assemblages and relations among assemblages of codes and norms – that establish control over persons and material objects. This conclusion does not, notice, impugn Giddens's insistence on the centrality of power in social life. It shows instead that resources are not ontologically "on a level" with rules, and that power ultimately rests on rules, not resources. In any event, the two fundamental dimensions of practice are the semantic and the normative. In evaluating Giddens's account of practice organization, consequently, we need attend to rules alone.

Recall that a rule is a methodically applied generalizable procedure of action implicated in the practical activities of daily life. In prescinding from treating rules as explicit formulations, this definition evidences considerable practical sensitivity: Rules are present in practices only in the practical consciousness through which actors participate in practices. Understanding a rule, moreover, is the same as knowing "how to go on," and a rule itself is something like a way of going on. Giddens, accordingly, characterizes practical consciousness, the understanding of rules, as a "generalized capacity to respond to and influence an indeterminate range of social circumstances" (CS, p. 21).

Since actors cannot articulate practical consciousness, Giddens might be read as suggesting that rules cannot be formulated. He does not, however, stake this claim. He instead advocates the considerably weaker thesis, that formulations are "already" interpretations of rules (CS, p. 23). This thesis does not imply that rules cannot be made explicit. It signifies only that formulations are a type of rule distinct from other types, for example, the unformulated ones embedded in practical consciousness.

Giddens, as noted, describes unformulated rules as "formulas" and "typified schemes" "invoked" in the course of day-to-day activities (CS, pp. 21–22). On his analysis, formulated rules differ from unformulated ones in the way formulations such as "The rule defining checkmate in chess is . . ." and "It is a rule that all workers must clock in at 8.00 A.M." differ from formulas such as "$a_n = n^2 + n - 1$" (CS, pp. 19–21). Formulated rules are explicit statements of rules in language, whereas unformulated ones are formulas tacitly grasped and applied by actors. Part of the point of saying that the formulas are "tacitly" grasped and applied, moreover, is to emphasize that understanding a formula is not the same as either uttering or being able to utter it.

Although there is a difference between a formula and a sentence, both are formulations. So the difference between unformulated and formulated rules for Giddens is the difference between tacitly grasped formulations of one sort and explicitly stated formulations of another. Rules, consequently, are formulations functioning in an implicit state. That is, what actors know in tacitly knowing a rule is what they would know if the formula involved was explicitly stated (and they still understood how to apply it). As a result, the unformulability of practical consciousness is simply actors' inability to formulate the formulas they constantly apply in daily life. Despite his many homages to practical consciousness, consequently, Giddens ascribes to it significant propositional elements. Practical consciousness, knowing how to go on, is grasping a formulation, knowing how to apply it.

As we have seen, practical understanding/consciousness cannot be formulated. Knowing how to go on is the mastery of a technique, of a way of acting, that in principle defies adequate formulation in words, symbols, diagrams, or pictures. This fact, in turn, problematizes the claim that practical consciousness is grasping formulas. To begin with, recall Wittgenstein's demonstration that formulations are unable all by themselves, that is, in the absence of established ways of following/applying them, to fix determinately what people do in following them. His observation in this context, that words, symbols, diagrams, and pictures, taken by themselves, can be systematically "followed" in indefinitely many ways (*PI*, 86, 139–41), shows that knowing how to go on can be modeled as understanding any formula you please so long as *how* people apply/follow it is suitably "adjusted" to match what they actually do. Even when the range of formulas that people are hypothesized to have been applying is restricted to prima facie plausible formulations, Wittgenstein's discussion of defining terms of natural language (see Chapter 2) intimates that every candidate formulation will fail to delimit practical understanding adequately. If, consequently, people are continuously applying formulas in daily life, it is impossible in principle to state the formulas – it is not merely the case, as Giddens has it, that people are contingently unable to do this. Formulas, however, that in principle cannot be stated are indeterminate, without specific content. So if people are constantly applying formulas, they are not applying anything specific. I doubt that the notions of unformulable formulas and of applying such formulas are intelligible. What is clear is that they are obscure and unenlightening. To avoid explicating knowing how to go on as the application of unformulable formulas, consequently, it must not be treated as the grasp of any sort of formulation. Practical consciousness embraces the grasp of formulations only to the extent that explicit formulations happen to circulate. (And

Giddens's ultimate mistake is thus to liken a procedure of action to a formula.)

Recall, incidentally, Giddens's contention that actors' inability to formulate rules is overcome by social scientists' ability to infer them as "recursively implicated" in regularities of practice (*CP*, p. 65). This supposition founders on the essential unformulability of practical understanding. Social scientific descriptions that recoup the regularities of a practice do not comprehensively demarcate *ante eventum* the further actions that could be performed on the basis of the practical understanding(s) organizing the practice. As a matter of logic, moreover, any finite set of regularities can be captured in indefinitely many sets of "rules." Social scientists, consequently, are in no better position than actors to formulate practical understanding.

It is also clear now why Giddens shrinks back from Wittgenstein's sheer practical sensibility. On his account, central to the analysis of a broad range of social phenomena such as institutions, systems, social reproduction, contradiction, power, ideology, space, and social change is a grasp of the rules and resources that structure practices. His outline of an analysis of social interaction similarly portrays this enterprise as relying on the discernment of these structural elements. The unformulability of practical understanding, therefore, denies these phenomena determinant content and renders social investigation of them inherently inadequate. Indeed, Wittgenstein's considerations problematize any account of social life that promotes rules as the chief determinant of action. By collapsing the organization of practices into practical understanding, consequently, Giddens, like Bourdieu, backs himself into an impossible situation.

Individual-Practice Homologies

Of the three accounts, Bourdieu's advances the strongest conception of a "homology" between individuals and practices. Narrowing the sites of consonance to practical understanding alone, he renders habitus and practices strict homologues in hypothesizing that the same system of opposition structures both. In relatively homogeneous, "traditional" societies, the fit between practices (conditions) and habitus approaches perfection and can perpetuate itself indefinitely. In more complex and heterogenous, "modern" societies, the proliferation, internal differentiation, and uneven development of fields and practices therein induce mismatches and time-lags between the dispositions of particular individuals and the organizations of particular practices. If no longer perfectly harmonious, habitus and practices still remain broadly homologous in being structured by the same oppositions. "The *habitus* is a metaphor of the world of objects, which is itself an endless circle of metaphors that mirror each other *ad infinitum*" (*LP*, p. 77; cf. *OT*, p. 91).

Giddens likewise confines the individual-practice homology to practical understanding. Unlike in Bourdieu, however, practices and practical understanding are not structured the same. They are in a sense identical. Rules and resources are, simultaneously, elements actors draw on when interacting and the medium of the spatiotemporal extension of practices. Thus, "Structure has no existence independent of the knowledge that agents have about what they do in their day-to-day activity" (*CS*, p. 26); and "Structure enters simultaneously into the constitution of the agent and social practices, and 'exists' in the generating moment of this constitution" (*CP*, p. 5). Although not all aspects of the individual (i.e., wants and reasons) are homologous with practices, the central dimension of practical understanding is at once the organizing medium of practices.

As discussed, Bourdieu's and Giddens's overloading of practical understanding engenders difficulties. Parallels, homologies, and identities between individuals and practices, as a result, should not be confined to this dimension of mind/action. The previous chapter maintained that understandings, rules, and teleoaffective structures organize practices. Knowing how to go on, following rules, and combinations of life conditions all represent sites for "homologies" between participants and practices. In the dimension of understanding, the essential identity that Giddens envisions between individuals and practices reigns. A person cannot participate in a practice without relying on understandings that at the same time organize the practice. Vis-à-vis explicit rules, on the other hand, the possibility that participants might contravene rules – and not merely disagree with while submitting to them (see *CP*, p. 102, on value standards) – implies that the identity Giddens posits holds only contingently if at all. Whether the rules that organize a practice govern a given participant's behavior is an open matter. In the dimension of teleoaffectivity, finally, the homology between individual and practice is not only contingent, but also not a matter of identity. It instead consists in the conformity of the teleoaffective order governing a participant's action to the teleoaffective structure of the practice. "Homology" here consists in a given action-governing order realizing one of the practice's normative possibilities.

Fields of Possible Action

Bourdieu and Giddens agree not only that a single structure governs actions and organizes practices, but also that practices open fields of possibility for their participants. (The idea that social phenomena establish such fields is widespread across social thought.) Despite this accord, uncertainties remain about the type of possibility opened and the value of conceiving practices thusly. Practice theorists almost always construe the possibilities established by practices as possible actions. My characterization of a practice's teleoaffective structure as a normative field of teleologies and affectivities is an anomaly in this regard. Bourdieu, for instance, writes:

> In short, being the product of a particular class of objective regularities, the *habitus* tends to generate all the "reasonable," "common-sense," behaviors (and only those) which are possible within the limits of these regularities. (*LP*, p. 55)

Similarly, Lyotard writes:

> A phrase "happens." How can it be linked onto? By its rule, a genre of discourse supplies a set of possible phrases, each arising from some phrase regimen. Another genre of discourse supplies another set of other possible phrases.[18]

Neither Bourdieu nor Lyotard focuses, however, on this aspect of practices.

Giddens devotes greater attention to the delimitation of possible actions. He does so partly because his "structuralism" differs from previous versions in explicitly conceiving structure not merely as constraining, but as both constraining and enabling. Rules and resources, in addition to closing off, also open up possible actions.

Giddens identifies three sources of constraint (*CS*, pp. 174–179): physical features of the body and material environment, such as the "packing constraints" discussed by Torsten Hägerstrand;[19] power, whose constraining effects are usually experienced as sanctions; and structure. I am puzzled about the distinctiveness of the second source, since Giddens systematically relates power to domination and treats the latter as a dimension of structure (asymmetric resources). In any event, on Giddens's analysis physical states of affairs in conjunction with rules and resources limit the options open to actors. These same matters also enable actions, in the sense of making them possible. Physical strength, for instance, opens up a variety of possible actions, as do a telephone connection and a clearing in a forest. Control over land or over the organization of activities similarly makes certain courses of action possible, just as the possession of particular rights does so.

Giddens identifies several key determinants of fields of possible action. Since, however, resources collapse into rules and the sort of implicit rule he countenances does not exist, his account of these components is inadequate. He fails to note, moreover, the existence of different types of possible action. Two such types, logically and physically possible actions, will be put aside in what follows. Although the actions that practices make possible are noncontradictory and compatible with physical law, it is not as such that they are primarily pertinent to behavior. Giddens works with a notion of possible action that coordinates possibilities with the structure of action intelligibility. Before considering this, it will be useful to examine "general" possibilities and how practices establish them.

A way of acting or being is "generally possible" for, or "generally open" to, someone when it is something that he or she is able to do or be. A person is able to be and do something, however, in several regards. A person can, first, be physically able to do something, say, play catch, depending on his physical constitution, the disposition of his limbs (they are not tied down), and physical features of the setting such as the shape of the built environment and physical properties of objects. A person can also be able to X in the sense of knowing how to do so. Physically being able to play catch is not the same as knowing how to do so. A person can, in addition, be socially able to do something in the sense that others' actions facilitate and/or do not prevent him from doing it – for example, he is given a ball, and others play along or do not stop him. An action is generally possible for someone, then, when he or she is at least physically, "skillingly," and socially able to do it. A person's general possibilities thus depend at least on physical states of affairs, know-how, and the behavior of others – thus, on materiality, knowledge and abilities, rules, teleology, and affectivity.

The provision of general possibilities is an important feature of practices. In participating in practices, people acquire knowledge and abilities, become cognizant of rules, build and alter the physical environment, and have their reactions and the teleoaffectivities governing them shaped and calibrated. Practices thus conspire with physical states of affairs to delimit what people are generally able to do – by outfitting people with the wherewithal to carry out particular activities, by helping to establish both what customarily makes sense to people to do and what is correct, prescribed, and acceptable in general, and by physically excluding some possibilities while admitting others. In American farming practices, for example, people learn how to operate threshers, milk cows, judge impending weather, build barns, plant grain, and gauge required quantities of irrigation water; they help one another with harvesting, uproot encroaching fence lines, gather at the general store, report those who mistreat animals, and keep close tabs on the amount of irrigation water others use in their fields; different crop combinations, sequences of spring planting tasks, uses of fertilizer, ends (monetary and life-style), planting and harvesting techniques, and the like are accepted or prescribed; and along with fields being laid out, barns built, trees planted, and fences erected, soils have certain properties, the weather is the weather, pests occasionally invade, and stores and markets lie at certain physical distances. Because of such matters as these, a range of ways of acting and being is generally open to an American farmer.

The ranges of general possibility that a practice establishes for its participants are not always well defined. To begin with, possible actions can be endless. What a person knows how and is physically and socially able to do within a practice is considerable. *By themselves,* in other words, know-how, materiality, and others' behavior leave open numerous possibilities. The spacious range of actions generally possible for an actor can be narrowed only by noting further mental and cognitive conditions of hers as well as additional features of her situations. Doing this, however, *eo ipso* means considering a type of possible action other than those generally open to someone.

A range of general possibilities can also be indeterminate. How others act is open, strictly speaking, until they do so. Given an actor's knowledge and abilities in conjunction with the likely (customary and normativized) behavior of others, moreover, it might not be definite whether a given action is in fact possible for her. And which novel and innovative projects and ways of acting are made possible by existing material structures and practices is indeterminate until they are realized or conceived in future action and thought. In sum, the range of actions generally possible for someone can be sufficiently complex, and is always sufficiently indefinite, to resist easy surveyance and presentation. Although practices open up and delimit the fields of action generally possible for people, representing these fields adequately is a tricky and at times intractable proposition.

In addition, how the factors delimiting general possibilities accomplish this is fractured. Physical states of affairs constrain action by excluding certain behaviors absolutely (until the state is altered). Many physical states also enable actions by putting something at a person's disposal that is needed in order to perform them (like a telephone connection, which makes telephoning possible, or arm strength, which makes winning the arm wrestling championship possible). Knowledge and abilities, on the other hand, do not (in a direct sense) exclude anything; only their lack renders a particular action impossible. Knowledge and abilities instead open possibilities by underwriting capacities to act. Normativity, meanwhile, excludes only in the sense of threatening transgressors with sanctions; and unlike both physical impossibility and the lack of knowledge/abilities, the threat of sanctions does not close off transgressive actions. The enabling effect of normativity, moreover, is the mirror image of its exclusionary power. Social approval, together with the absence of threatened sanction and approbation, does not so much make actions possible as make them easier. In one sense, therefore, normativity does not help circumscribe the actions generally possible for people. It does so in another sense, however, because in being a generic feature of situations of action, it opens certain higher-order possibilities (e.g., those of conforming unthinkingly to the norms or of shirking and disrupting them).

Finally, the actions of others delimit a person's general possibilities by marking actions feasible and infeasible. This form of enabling and constraining is unstable and potentially fragile: Indeterminacy always attends how others will act in a given situation; dialogue (and inducements) can often change what they do; and even in situations where dialogue (and inducements) are ineffectual or excluded, it might be possible to circumvent how their actions seem to render a way of proceeding infeasible. Others' actions reliably and stably constrain as well as enable a given actor's actions only to the extent that normativity has so impacted intelligibility that people automatically conform to custom and to what is otherwise correct or explicitly prescribed.

So a person faces a field of general possibilities in the senses that (1) extant physical structures exclude certain actions and make others physically possible, (2) he or she possesses the knowledges and abilities required to perform certain actions, and (3) she or he participates in practices in which certain actions are normatively favored or sanctioned and people tend to act in certain ways. This is neither a unified nor necessarily a very restrictive constellation of delimitations. Once again, therefore, fields of general possibility prove themselves not the most accommodating object of study.

The possibilities that Giddens alleges materiality and structure delimit resemble general possibilities, but are further circumscribed. Their nature is revealed by his occasional remark, that structure enables and constrains the *feasible options* open to actors in given circumstances (e.g., *CS*, p. 177). Feasible options are narrower in scope than general possibilities. Although generally possible actions are in some sense both feasible and options, "feasible options" are ways of sensibly proceeding in a given context. That an action is possible in this sense is tied to action intelligibility. Feasible options closely resemble what Dreyfus calls "existential possibilities." As he explains, existential possibilities are, in William James's words, "live options": ways of proceeding that someone just might take up in a given context.[20] Actions are live, or feasible, options for someone in a particular context when they are generally possible for that person and fit what he or she is up to in that context. They are the ways of proceeding that are sensible in that context from the perspective of the actor.

Giddens and Dreyfus imply that an actor usually enjoys a range of feasible options, or existential possibilities, in a given context. Before

considering the sense in which this is true, note first that it would be
not be true if "sensible way of proceeding" meant what is signified to
an actor to do. As discussed, signifying is the specification of what
now makes sense to someone to do. It does not, however, identify a
range of actions that now make sense. Like Bourdieu's habitus, it
instead "selects" one action as the one to perform. If signifying did
not pick out a specific action, nothing in the realm of intelligibility
would close the resulting gap between sensible possibilities and what
the person actually does. Why someone performed one sensible option rather than another would be inexplicable, for there would be
nothing with which to explain it, even ex post facto. One could only
look around, possibly in vain, for a causal explanation. (Expressed
in Giddens's language, the idea is that given materiality, rules and
resources, the disposition of others, and all a person's wants and
reasons operative at a given moment, there could have been only one
thing for her to do at that moment.) At a given moment in a particular context, therefore, it is not the case that a range of sensible ways
of proceeding is signified to an actor. Indeed, at any such moment
he or she is always already doing something particular.

Giddens suggests that a person's feasible options depend on her
wants, more generally (it can be supposed) on her conditions of life.
He notes, for instance, that the truth in Marx's claim, that under
capitalism workers must sell their labor power to employers, is that
workers have only one feasible option given their wish to survive (*CS*,
p. 177; see also the discussion of compliance at *CS*, p. 309). Giddens's
point about the dependency of options on life conditions is important
and correct; his application of it to capitalist labor, however, is incomplete. Selling their labor power is the single feasible option for workers not merely given their desire to survive, but only given further
wants and conditions as well, for example, the desire not to be, or
the known lack of knowledge required to become, a government
bureaucrat. In fact, workers have just one feasible option only given
all those wants, conditions, circumstances, and, moreover, know-hows
(rules and resources) of theirs relevant to earning a livelihood. Indeed, the range of feasible options that anyone enjoys in any context
whatsoever is relative to all those wants, further conditions, and
components of practical consciousness of hers that are relevant to
what she is up to in that context.

In Giddens's example, what workers are taken to be up to in the
context of capitalism is earning a livelihood alone. At any particular
moment in a given context, however, what someone is up to is as
multiple as the teleology governing his behavior is jointed. As discussed, most actions are performed for the sake of some end, maybe
also as part of some project for the sake of that end, maybe also as

part of some task that is a stage in a project which is carried out for the sake of that end, and so on. At any moment in a given context, consequently, what a person is up to might encompass pursuing a given end, carrying out a project for the sake of that end, and performing a given task as part of that project. Given normativity, circumstances, and the actor's other conditions (including moods and emotions), there might also be a range of projects that can be sensibly performed for the sake of that end, a range of tasks sensibly carried out as stages in that project, and so on. Of course, this might not be the case. No other project might make sense for the sake of that end (as with laborers under capitalism on Marx's analysis) and no other task as a stage in that project. If, however, alternatives do exist, a person enjoys a range of feasible options, or existential possibilities. Indeed, he just might enjoy several such fields, each associated with a different "level" of the teleological order that structures what he is up to. What is more, a person might also at a particular moment in a given context be prepared to act for different ends and thus to carry out hierarchies of projects, tasks, and actions that make sense for the sake of those ends. If so, he faces additional arrays of fields of feasible option, each field coordinated with a different level of a teleological order that could come into play if he were to begin acting for the sake of something different. In sum, an actor almost never faces a single field of existential possibility, unless we treat the sum of the arrays of smaller fields as a single field.

Such fields are not merely multiple, or at least extremely variegated. More important, they are abstract constructions irrelevant to the flow of unreflective behavior. Unreflective behavior is governed by signifying, which specifies one particular action at a time. As long as the flow of reactions persists, the signifying of particular actions continues uninterrupted. What is signified, moreover, is determined by particular teleoaffective orders, by particular constellations of ways in which things stand and are going for the actor. Fields of existential possibility, by contrast, are abstract entities that are constructed by holding some components of the structure of signifying (ends and projects, etc.) constant and seeking to replace other components with alternatives. Such fields inform their constructors of possible variation in signifying, but do not determine what actually is signified.

They do not even "govern" behavior in the sense of constraining and enabling signifying. What accomplishes this is, first, the collection of ends for the sake of which someone is willing to act and, second, the range of projects, tasks, and actions with which he is familiar. The latter range delimits the possible components of signifying chains. Signifying is further enabled and constrained by the particulars of the circumstances as registered in the actor's beliefs, expectations, and the like; by the different emotions and moods in which he might find himself; and, in a different manner, by the normativity of understandings and teleoaffective structures (e.g., how entities can be acceptably and correctly acted toward, and what can be acceptably or correctly pursued or done for the sake of or as part of X).

A field of existential possibility constrains and enables nothing. It simply registers the range of sensible ways of proceeding that results from treating certain ends, projects, and tasks as fixed and others as substitutable, given a range of moods and emotions, familiar actions and projects, practice organizations, and understandings, beliefs, and expectations concerning circumstances. That such arrays of possibility constrain and enable little is suggested by the fact that a person can suddenly shift from acting for one end to acting for another, or from carrying out one project for a given end to carrying out another, and so on. Until action occurs, it is never determinate which end a person will have acted for, what project he will have carried out for that end, what emotions will have affected this, and even whether he will have acted for any end at all.

Fields of existential possibility help determine action only when an actor thinks about what to do. Practical deliberators often ponder which actions make sense given certain ends, projects, practices, and circumstances. They also occasionally construct arrays of sensible action so as better to decide which to pursue. These are, however, the only sorts of situation in which fields of existential possibility help determine action. *Unreflective* action is always governed by a signifying that "selects" a single action and neither delineates such fields nor is constrained by them. Incidentally, none of this denies that insofar as ends, emotions, circumstances, and practices are stable in social life, the construction of fields of existential possibility can be a propitious model-building tool in social science. Investigators' representations of such fields might even be more accurate and comprehensive than those constructed by practical deliberators.

I wrote earlier that practices establish fields of general possibility by establishing constellations of physical phenomena, outfitting actors with knowledge and abilities, exhibiting teleoaffective structures, and laying down rules and customary ways of acting. They also therewith constrain/enable signifying as well as open fields of existential possibility by providing materials – circumstances, normativities, understandings, and familiar ends, projects, and actions – that constrain/enable signifying and with which investigators and practical deliberators can construct such fields. Together with the physical environment, therefore, the total nexus of extant and past practices delimits the outer horizon of both general and existential possibilities. This means, among other things, that there is no carrier of these possibilities wider than networks of practices. Characterizing cultures and societies as establishing or as themselves fields of possible action requires analyzing cultures and societies by reference to networks of practices and their organizations.

All in all, fields of possible action have mixed value for the investigation of social life. Since the fields of general possibility established by practices are indefinite, typically complex, and heterogeneously determined, they cannot be easily represented. We cannot, therefore, expect social inquiry to provide detailed descriptions of them. Fields of existential possibility/feasible option can, by contrast, be constructed, although their greater simplicity and determinateness in comparison to fields of general possibility often reflects the penchant of social scientists and practical deliberators to abridge the complexity of circumstances, the depth of teleology and affectivity, and the breadth of familiarity. These constructed fields, however, are usually irrelevant to actual life. Social science will undoubtedly continue to find it useful to construct fields of feasible option, for instance, the range of "rational" options reflecting others' likely behavior. The usefulness of such devices for investigatory purposes, however, must be distinguished from their considerably slimmer role in social life itself.

6

Practices and Sociality

By itself, the practice theoretical consolidation of practice as nexus and practice as do-ing reveals practices as a site of sociality in human life. For the interrelatedness of participants in a practice is secured merely by the fact that the understandings, rules, and teleoaffective structure organizing the practice govern actions of *all* participants. As we shall see, practices open fields of sociality and social order richer than simple commonalities in what governs behavior. These more elaborate fields also furnish the material with which the composition of social formations such as groups, institutions, and systems can be analyzed.

The current chapter stops short of substantiating the latter thesis by carrying out the indicated analysis. It instead simply makes good the claim announced in the opening chapter, that an account of practices can be developed on the background of Wittgenstein's late remarks on mind and language (1) that reinforces the practice theoretical faith that practices represent an alternative starting point for social ontology; (2) that offers a platform for the ontological comprehension of social affairs that is broader and more adequate than those represented by current accounts of practice; (3) that respects the locality and complexity of social affairs emphasized not only by practice theorists, but also by thinkers contentiously labeled "postmodern"; and (4) that delivers additional insights into the social constitution of the individual, thus reinforcing the refusal to make individuals the exclusive starting point for the comprehension of social life. Previous chapters have both argued for the superiority of my account of practice over those of Bourdieu, Giddens, and Lyotard and explicated how practices on this account constitute individuals. The current chapter motivates the ontological centrality of practices in social life and sets an agenda for further theorizing in this direction.

My path toward these goals begins by explicating and criticizing two traditional approaches to the nature of sociality. Both approaches compose sociality out of individuals and relations among individuals *simpliciter*. I argue that the individualist material out of which they construct sociality implicates practices and thus the more complex sociality opened there. The advent and persistence of individuals, by virtue of occurring within and on the background of practices, is *eo ipso* the institution of subjects who are integrated into practice sociality. Following this, I describe several key components of the tissues of sociality immanent in practices. This explication concretizes the claim enunciated in Chapter 4, that practices are inherently social because participating in one entails entering a complex state of coexistence with other participants. In the latter parts of the chapter, finally, I sketch in broad strokes the social ontology that results from treating social life as a nexus of practices and the sociality opened in this nexus as the basis, or "substance," of all sociality in human life. This sketch lends contour to the thesis, that the composition of such social phenomena as groups, institutions, and power can be analyzed via practice sociality. My comments in this final phase are, however, schematic. Following through with this sketch, detailing how social formations and phenomena can be analyzed in terms of practice nexuses, is an immense task whose proper execution lies beyond the scope of this book. The current work aims primarily to develop the notion of practice that can figure in an adequate account of the social field as a nexus of practices; and the course of the current chapter leads only to the frontier of this practice theoretical ontology and offers hand-waving characterizations of its topology.

Before beginning, I should briefly defend the focus on individual practices adopted in this chapter prior to its concluding section. Practices, as already noted and will be further discussed in that section, are invariably related. The socialities they open are similarly linked and interdependent. These facts do not, however, nullify the inherent "socialness" of individual practices, taken by themselves. Furthermore, the socialities opened in individual practices exhibit general properties that are independent of the facts that practices are interconnected and their socialities interwoven. Analysis of these general properties, consequently, does not presuppose and proceed from prior investigation of general features of both practice nexuses and the socialities they open. *Pace* the intuitions of systems theorists (and others), inevitable contingent relatedness among individual practices does not entail the theoretical dependency of general properties of individual practices upon those of nexuses.

Sociality and Social Order Defined

It is best to begin by reminding the reader of the senses in which the expressions "sociality" and "social order" are employed here. "Social," I suggested in Chapter 1, means pertaining to (human) coexistence. A phenomenon is social, accordingly, when it pertains to human coexistence. Institutions, for example, are social not only or even firstly because they encompass scores of individuals, but even more so because they configure coexistence among individuals.

I emphasize that this definition of "social" does not prejudge the proper analysis of the social. In tying "social" to coexistence and rendering coexistence as that of individual human beings, the definition appears to steer subsequent analysis, if not in an individualistic direction, then at least counter to "holistic" or "systems" approaches. By an "individualist" analysis I mean one that construes social life as nothing more than individuals (i.e., their mind/actions or actions alone) along with relations among individuals. The impression of individualist bias is incorrect. The equation of social with coexistence does not require that coexistence be analyzed individualistically. It entails merely that a theory, to be a social theory, must be about matters pertaining to human coexistence. Many macrotheories, for instance, do not discuss individuals and their interrelations, but instead analyze constellations of such entities as roles, institutions, structures, and systems. These institutions, structures, and constellations qualify as social phenomena when, according to these theories, human coexistence is established and/or thrives within or as these institutions, structures, and constellations. This, in turn, is the case when the theories implicitly or explicitly bind human togetherness to these entities – that is, when their analyses either imply or narrate that human coexistence takes certain forms because of institutions, structures, and constellations and how they hitch to individuals. Lack of analysis of the coexistence per se of individuals, in conjunction with a preponderance of attention to the characteristics of institutions, structures, systems, and constellations thereof, does not mean that these theories fail implicitly to render human coexistence as beholden to these entities. I am not suggesting that the aims and cruxes of macrotheories are always best captured by highlighting how the institutions, structures, and systems they theorize configure coexistence. Any theory, however, that either obliterates or makes opaque its implications for coexistence forfeits its title as a theory of social life.

Essentially every account claiming to be of social life passes this test and thus is compatible with, if not also an advocate of, the equation of social with coexistence. This even holds of the overwhelming majority of systems theories, for instance, those which conceive systems as systems of interactions among individuals.[1] Indeed, the pride of place accorded to interactions by many sociologists in their general accounts of social life confirms the widespread intuitive identification of social with coexistence.[2] What's more, all systems theories that construe systems as composed of actions but prescind from restricting the actions involved to interactions, implicitly reproduce the assimilation of social to coexistence. For any such account portrays systems, and therewith social life more generally, as linked actions (i.e., the linked actions of individuals), thus as a form of human "togetherness" in which coexistence consists primarily in an unintentional interrelatedness of people's actions. In Habermas, for example, the economy is a system of success-oriented purposive-rational actions coordinated in a nonconsensual manner over money, which differentiates out of the totality of actions composing society in fulfillment of that function of the totality that Parsons called "adaption" (reproduction of the material substrate). The coordination of different individuals' actions over money is, of course, a form of coexistence. The economy, consequently, is a large-scale state of human coexistence encompassing a vast and open-ended range of individuals (i.e., actions).[3] In sum, the identification of "social" with human coexistence does not favor any particular social ontology. It leaves open the proper analysis of human coexistence and designates this a prize fought over by the different stripes of individualism, wholism, and practice theory.

Coexistence, moreover, can be further characterized (again neutrally) as the hanging-together of human lives,[4] as the existence of *Zusammenhänge* among people. As indicated in Chapter 1, a *Zusammenhang* is a nexus of phenomena that forms a context for each phenomenon involved. Phenomena form a *Zusammenhang* when they hang together and, as so joined, form a context within which each is and proceeds. Human lives form a *Zusammenhang*, consequently, when people's mind/actions hang together and, along with their identities, subjecthoods, genders, and person types, evolve in a manner reflecting this dependency. When the placeholder "hang together" is unspecified, so that it remains open whether the hanging-together of lives is a matter solely of relations among lives, this formulation is sufficiently vague as to encompass the great variety of past and extant accounts of social life – from extreme individualist analyses such as Max Weber's confinement of sociality within the contents of individuals' mental and cognitive conditions, and system theoretic conceptions of the establishment of *Zusammenhänge* through symbolic media of communication, to practice theoretical concepts of integrated understandings such as Bourdieu's notion of homologous habitus.

"Sociality," then, designates the dimension of formative context-constituting coexistence in human life. An account of sociality, accordingly, is an account of coexistence, and a form of sociality a form of coexistence. The "social field," moreover, is the overall site specific to coexistence where coexistence transpires. (I write "specific to coexistence" so as to disqualify such "sites" as the planet Earth and the solar system.)

"Social order," furthermore, means in the current context the ordering of lives. An "ordering" can be further specified as an arrangement of individuals in which each has meaning and place. Although this notion of social order diverges from the more familiar notion of regulated and regularized nonovertly violent coexistence, it remains neutral over against competing general approaches to sociality. Any state of sociality, moreover, embraces a social order(ing), for any state of contextualizing coexistence necessarily implicates an arrangement of individuals in which each has place and is positioned with respect to the others. Social order is thus an omnipresent feature of sociality that indicates something, as we shall see, about the "distribution" of human lives.

It follows from these definitions that the thesis, that a practice opens a field of sociality, avers that a practice establishes a tissue of coexistence among its participants that arranges them vis-à-vis one another and molds the progression of their lives (and identities) within the practice. The accompanying theorem, that practices are the locus of sociality in human life, embraces two subclaims. The first is that any attempt to build sociality out of individuals and relations among individuals alone founders on the fact that the individuals and relations involved exist only within practices. Individuals, to speak with Heidegger, are "always already" caught up in the tissues of arranging-contextualizing coexistence established in practices. That is to say, there is no individual who does not *eo ipso* coexist with others in the medium of practices. The second subclaim asserts that the sociality of practices is that out of which social formations such as families, artistic movements, governments, and economies are composed. The multilayered socialities that characterize such formations are built out of the forms of coexistence opened in practices. Hence, the sociality of practices is fundamental in human existence in contextualizing individuals essentially and underlying social formations and other social phenomena (such as social structure) ontologically. This, then, is the full sense of the claim that practices are the site where human coexistence is established and ordered: All dimensions of human coexistence ultimately refer to practices.

Although these claims concentrate sociality within the province of a particular type of entity, this regimen has antitotalizing implications. Common to practice theories is a picture of the social as a complex nexus of practices. These nexuses are not large-scale totalities that possess features beyond those of the aggregate of their constituent practices and observe dynamic and systemic principles that apply to them *qua* wholes. Although nexuses of practice reveal general properties to the theoretical gaze, these properties are characteristics of manifolds, not traits of totalities. As mentioned in Chapter 4, moreover, practices vary in different locales. Dispersed practices such as explaining, questioning, and ordering and integrative practices such as those of farming, business, and cooking vary across history or social space. So the picture of the social field as a nexus of integrative and dispersed practices is a flexible portrayal that, in addition to resisting totalization, endorses complexity, difference, and particularity.

This picture of the social respects complexity and particularity in a second way. I wrote in Chapter 1 that my account of sociality – at present, the picture of the social field as a nexus of integrative and dispersed practices – does not claim completeness as an account of social life. In laying out the basic nature of the social, it instead constitutes a framework through which to investigate social domains and phenomena and uncover their local details and complexities. For instance, conceiving of social formations such as families and economic systems as "slices" of the nexus of practices that composes the social field (see the last section) provides a lens through which to look at particular formations and to discover the details within them of the *Zusammenhang* of lives. Conceiving of the social field and of social formations thusly says little about these details. It simply implies first, that life, interaction, and coexistence are local and particular "processes" that occur within the nexus of practices generally and specific slices thereof more particularly; and, second, that particular practices – their organizations and connections – thus form the central and omnipresent context in which these processes occur.

Because the sociality of the human sojourn is established in integrative and dispersed practices, it is laid down in the specific and intricate coexistences opened in nexuses thereof. The variableness and complexity of these coexistences ensure that sociality writ large is a vast *Zusammenhang* of lives that modulates across social space and time, is rooted in the density of local circumstances, and never collects into or makes up something more than the contingent and shifting linkages of the variable and complex arrangements involved. Human coexistence has no further scope or more systematic countenance than those displayed by the nexuses of practices in which it is constituted.

Two Classical Approaches to Sociality

I will work toward my account of practice sociality by first reviewing two prominent and well-known approaches to the nature of coexistence. These two approaches do not, of course, exhaust the going perspectives on sociality. As the previous section insisted, all social ontologies either explicitly or implicitly analyze the nature of the arrangement-establishing *Zusammenhang* of human lives. Differences among ontologies in this regard represent disagreements over both that through or on the basis of which lives hang together and the character of the configurations that thereby result. Only some traditions of social thought, however, have carefully and directly scrutinized the nature of *Zusammenhänge* and that through which lives hang together; and most of these reduce *Zusammenhänge* to individuals and relations among individuals *simpliciter*. This situation is not surprising. Human *Zusammenhänge* are an explicit topic primarily for approaches focusing on individuals because the linking of individual lives is a more salient phenomena for these approaches than for those which compose the social out of entities other than individuals (e.g., institutions or subsystems).

The first of the approaches I will consider, most consequently pursued in phenomenology, examines sociality from the perspective of the individual. What I mean is that it construes sociality as a matter of how an individual encounters other people in lived experience mediated by understanding. Alfred Schütz offers perspicuous examples of this sort of analysis. For example, his early *Hauptwerk, Der sinnhafte Aufbau der sozialen Welt,* begins with an analysis of the constitution of meaning and meaningful experience in the mental life of an isolated consciousness. The book's analysis of the social world only begins subsequently, with the question of how an I can understand others. Schütz abandons a Husserlian framework, which would treat this question as concerning how the other is constituted within the subjectivity of an isolated I. The issue nonetheless remains for him how in everyday life an individual I, or person, grasps other people's streams of experiences and intended meanings (*gemeinten Sinne*), which the I assumes to exist and to exist outside his own consciousness. For Schütz, consequently, to analyze sociality is to examine how an I encounters, that is, experiences-comprehends, others (their streams of lived-interpreted experience). It follows that the different dimensions ("regions") of sociality are demarcated by reference to features and possibilities of this encounter.

In line with the presumption, that the presence of a You in the I's experience qualifies subsequent analysis of that experience as of sociality, Schütz begins his account of the different forms of sociality by defining the "social world" as the world of the individual I populated by others (*die von Mitmenschen belebte Welt des einzelnen Ich*). He then divides this world into regions by reference to different ways others are given to the I and the particular techniques through which the I can understand them.[5] The "surrounding social world" (*soziale Umwelt*), for instance, is composed of those others (and artifacts and settings) who, in spatially and temporally coexisting with the I, are accessible to the I in direct experience. Since in this world others are "bodily" present, the I can immediately comprehend their streams of lived-interpreted experience on the basis of experientially apprehending their bodily movements and expressions. The "social with-world" (*soziale Mitwelt*), moreover, consists of all those contemporaries (and artifacts and settings) who, spatially absent, cannot be directly experienced by the I at a given moment, but could be directly experienced given proper changes in the spatial locations of the individuals involved. Comprehension of occupants of this region of the social world cannot transpire in direct experience of them and relies instead on indirect indices.

In this early work, sociality reduces for Schütz to a series of I–you, I–we, and I–them relations; and different regions of coexistence are cataloged by reference to the modalities of this encounter, the various whats and hows of the I's acting-understanding directedness (*lebendige Intentionalität*) toward others' interpreted-conscious experiences. As a result, the arrangement-setting *Zusammenhang* constitutive of sociality is above all the setting of individuals into different regions of the social world on the basis of the character of their experience and comprehension of one another.

When we speak here of the surrounding world [*Umwelt*], prior world [*Vorwelt*], with-world [*Mitwelt*] and ensuing world [*Folgewelt*], it is of course only meant that others are fellows [*Mitmenschen*], compatriots [*Nebenmenschen*], predecessors, and successors for me, that I myself am fellow, compatriot, predecessor, and successor for others, that I am able to look toward the consciousness experiences of these others, always in a particular way, just as these others are able to look toward mine.[6]

In later works, above all *Structures of the Lifeworld*,[7] Schütz enriches his analysis of the ordering aspect of sociality by incorporating into his account of an individual's understanding-experience of others the origin and distribution among individuals of the knowledge on the basis of which they experience and understand one another. This allows him to diversify the analysis of the relations among individuals that are constituted by their understanding-experiences of one another and thereby deepen his account of the *Zusammenhang* of lives. Although this addendum departs further from the subjectifying Husserlian framework from which Schütz took his initial orientation, it retains the principle that human coexistence is above all a matter of individuals encountering one another.

The living encounter of individuals with one another is an important feature of social existence, especially of those arenas of social existence embracing human interaction. At the same time, however perspicuous the analyses of Schütz and others might be of such experience, encountering does not exhaust the constitutive media of human coexistence. Lives hang together not merely through people's understanding-experiences of one another, not merely by virtue of people being the objects of one another's perceptions, understandings, beliefs, intentions, and actions. Moreover, the further sorts of state of affairs through which lives hang together condition how people encounter one another. For all its excellence as an analysis of experience, Schütz's account remains beholden to a neo-Cartesian modus operandi that commences analysis of the world by examining how the world is manifest in conscious experience and bases subsequent analysis on the results of this initial investigation.[8]

A second traditional approach to sociality, which I will dub the "compositional" approach, had a long and influential history in social theory from the latter two decades of the nineteenth century up until World War II. In maintaining that human lives hang together by virtue of commonalities and freely chosen bonds, it construed sociality as a composition out of individuals. An orienting landmark in this tradition was Ferdinand Tönnies's distinction between *Gemeinschaft* and *Gesellschaft*.[9] His notion of a *Gemeinschaft* conceptually enshrined a form of life that was disappearing amidst the incisive transformations in human coexistence wrought through industrialization and modernization: a natural concord (*Eintracht*) of wills, in conjunction with reciprocal, binding sentiment (understanding), that takes the form of customs, shared beliefs, and a common life in possession and enjoyment of shared goods (paradigmatically: common dwelling and living spaces). The coordinately ascending form of life highlighted by modernizing, especially capitalist, transformations was society (*Gesellschaft*). Society is a union of rational wills brought about by individuals freely entering into agreements, fashioning conventions, and thereby directing and binding their actions with regard to specific affairs and arrangements. Two important differences between these general types of social formation are (1) that in a community, union (*Verbindung*) precedes association (*Bündnis*), whereas in a society the order of precedence is reversed; and (2) that membership in a community pervades most dimensions of a person's life, whereas membership in a society only impacts people's activities with regard to the agreed-upon arrangements involved. That *Gemeinschaft* and *Gesellschaft* are the two basic forms of social formation for Tönnies implies that in his eyes human coexistence amounts to lives hanging together through common will and sentiment as well as freely and mutually chosen bonds.

Tönnies's distinction impacted social thought decisively until the outbreak of World War II. Although with time the distinction became less one between two types of social formation and more one between two forms of human association (which can combine differently in different concrete social formations), and although compositionist conceptions of the nature of commonality varied, the basic division was incontrovertible. Max Weber's typology offers an excellent example of continuity through these changes. Weber stipulated that a social relation exists when a plurality of (*mehrere*) individuals are oriented in their actions by mutual attitudes toward one another (or when there exists the chance that they are so oriented). Tönnies's community and society become in Weber's hands communalization (*Vergemeinschaftung*) and societization (*Vergesellschaftung*), two forms of social relation among individuals. Social relations are communalized when the attitudes toward others that underlie an actor's orientation in action rest on a subjectively felt belongingness to the individuals involved; and they are societized when these attitudes rest on a rationally motivated accommodation or joining (*Verbindung*) of the interests of those concerned.[10] Weber's definitions illustrate the tenacity of Tönnies's contrast between commonality and free association, even though for Weber the commonality involved is a shared feeling *of* belongingness and not, as for Tönnies, shared wills and sentiments more generally. Indeed, Weber stresses that commonalities, including those of language, are insufficient in themselves to guarantee communalization. The community must be part of the object of the shared mental condition in which commonality consists. This change arises in part from Weber's extremely individualistic equation of human coexistence with individuals' orientations toward others. Given this narrowing of the locus of sociality, any fundamental contrast in modes of coexistence must be specified by reference to orientations toward others, in this case, through that on which these orientations rest. This change also reflects the widespread propensity of compositionists after Tönnies to identify the commonality that binds individuals into groups as the shared possession of mental and cognitive conditions that have we-ness, or togetherness, as object. Other examples of this predilection are Georg Simmel's contention that the unity defining a group amounts to nothing more than the group member's consciousness of unity, and Gerda Walther's thesis that a community is constituted by a shared feeling of belonging together.[11]

In this tradition, groups occupy pride of place because they are considered the locus of sociality in human life. Since, moreover, groups are collections of linked human beings, coexistence amounts primarily to the linkage of individuals into groups. Community and society, accordingly, are the two overarching categories of groups, and communalization and societization (or their equivalents) the two overarching forms of linkage among individuals. Compositionist texts, as a result, are replete with typologies of the different types of community and society together with discussions of the different modalities of communalization and societization.

I will not spell out these ideas further. The discussion so far suffices to put a second approach to sociality on the table, one that construes human coexistence as individuals joined by commonalities and freely entered-into bonds.[12] I point out in passing that the compositional tradition must be distinguished from a more extreme individualist way of thinking that sees *Vergesellschaftung* as the sole paradigmatic bond among individuals constitutive of human coexistence. A representative of this third tradition is Oakeshott.[13]

Both the approaches outlined here shrink sociality into the lives of individuals: Human coexistence is constituted through the experiences, minds, and actions of individuals. These two approaches thus exemplify a wider individualist way of thinking that remains extremely influential today in a number of garbs, including neoclassical economics, symbolic interactionism, and ethnomethodology. Most proponents of these schools of thought do not pursue ontological analysis. One exceptionally clear exception is Harold Blumer, who extends compositionism in writing that the "empirical social world" is

ongoing group life. . . . This world is the actual group life of human beings. It consists of what they experience and do, individually and collectively, as they engage in their respective forms of living; it covers the large complexes of interlaced activities that grow up as the actions of some spread out to affect the actions of others, and it embodies the large variety of relations between the participants. . . . The life of human society consists necessarily of an ongoing process of fitting together the activities of its members.[14]

As Blumer explains, the "ongoing process of fitting together the activities" of members is nothing but members adjusting their activities to those of one another. He asserts, for example, that society is a totality of "joint actions," which are collections of action performed by actors who fit their behavior together in the knowledge that their actions are part of a collective activity (pp. 70–71). Examples are wars, trading expeditions, games, court trials, and shopping forays. He also relates that

Symbolic interactionism sees these large societal organizations . . . as arrangements of people who are interlinked in their respective actions. . . . At any one point the participants are confronted by the organized activities of other people into which they have to fit their own acts. The concatenation of these actions taking place at different points constitutes the organization of the given molar unit. (p. 58)

For Blumer, the social world is a totality of coordinated actions that are performed by individuals who take account of what others do, especially those with whom they interact (pp. 70–71). It thus consists in an interrelating of individuals through experience/mind and actions.

I might point out that many proponents of these two "classical" approaches, in contrast to their economist, interactionist, and ethnomethodological descendants, privilege mind as the primary locus of sociality, though they identify different mental and cognitive conditions as the medium of coexistence: understanding-experiences in Schütz, wills and understanding in Tönnies, consciousness in Simmel, feelings in Walther, orientations in Weber, unconscious states in H. Schmalenbach, or some combination of these as in Alfred Vierkandt.[15] This tendency is particularly strong in their analyses of community and communalization. Most of these theoreticians grant, of course, that sociality also consists in actions, specifically interactions or reciprocal actions (*Wechselwirkungen*), especially in the case of society and societization. But sometimes they acknowledge actions as a medium of human coexistence, even vis-à-vis society, only to the extent that actions are grounded in the mental states in which sociality is essentially rooted.

In any case, the basal motivation for these individual-*ansetzende* forms of analysis is obvious. Individual human beings and their interactions are an overwhelmingly significant feature of experience. When, consequently, "social" is intuitively understood to mean pertaining to the coexistence of human beings, approaching the nature of the social through properties associated with individual persons enjoys a certain experiential obviousness and palpable convincingness. Of course, other experiential phenomena also carry great significance (e.g., animals and lightning flashes). But there is nothing else of comparable import in experience to which to refer sociality and coexistence.[16]

Earlier I criticized Schütz for confining human coexistence to encounters. Lives do not hang together merely through people encountering one another. Not only, however, is it more plausible that mind/action more broadly, together with the setting of action (see two sections hence), is the medium of coexistence; it is eminently likely that mind/action and setting are the media through which lives interrelate. That is to say, a strong case can be made that interrelations among people transpire solely through mind/actions and features of settings. (These two greater plausibilities help explain, incidentally, why I develop my account by reference to individualism alone.) If, therefore, human coexistence is a matter of interrelations among individuals *simpliciter*, the compositionists are on the right track in tackling sociality via the commonalities and free associations that link individuals.

But is coexistence a matter of individuals and interrelations alone? Missing from compositionist formulations is explicit recognition that the minds, experiences, and actions through which lives hang together occur only within, and depend somehow upon, a wider context. Compositionist theorists would not deny that individuals are embedded in such constitutive contexts as language and custom. But these truths are not incorporated into their definitions and typologies of social formations and forms of association. For them, human coexistence is a matter of groups, groups are collections of linked individuals, and the links consist primarily in shared mental and cognitive conditions along with reciprocal or interdependent actions. Sociality is thus a matter of nothing more than interrelations transpiring through mind and action. If, however, mental commonalities and reciprocal actions exist only in the medium of some further entity, the states of coexistence that they constitute also exist only within that entity. In such a case, this further entity, as much as interrelations, is an aspect of sociality, an essential component of the hanging-together of lives. Now, the combined import of Chapters 3 and 4 was in part that conditions of life – mental and cognitive conditions as well as actions – presuppose practices. It follows that interrelations among lives that transpire through life conditions also presuppose practices. Sociality, consequently, is not merely a hanging-together of lives through mind/action and setting, but such a hanging-together as established by and otherwise transpiring within practices. Sociality is essentially an interrelating of lives within practices.

Before arguing for and further developing this fuller delineation of the nature of human coexistence, I acknowledge that practice theory is not the only way of thinking that maintains that lives only hang together within some wider, contextualizing entity. Some forms of systems theory, for instance, maintain that actions are linked by generalized symbolic media of communication, which thus form a wider context for the hanging-together of human lives. Although the relevant systems theorists disagree in their specific conceptions, they agree that a generalized symbolic medium of communication is roughly an avenue over which actions of one person influence those of others. Parsons, for example, writes that

[s]uch media, like language, *control* behavior in the process of interaction. They do so, however, by *symbolic* means, i.e., by presenting the actor, not with an intrinsically important object, such as a food object, but with a symbolic "representation" of such an object.[17]

In addition to language, Parsons identifies four such media: money, power, value commitment, and influence. Early Luhmann explains the idea as follows. Conceive of action as a selection from possibilities (and thus as a reduction of complexity). The function of each medium is to transmit this selection, or reduced complexity, so as to restrict the selection possibilities of others.[18] Because such transmissions normally make actors' selections the basis of other actors' selections, the media are effective means of control and influence. According to Habermas, finally, what distinguishes the media of money and power from the medium of language is that each is a mechanism of action coordination operating independently of the processes of linguistic consensus building. They operate not by way of rationally motivated persuasion, he writes, but instead through empirically motivated ties (Bindungen), that is, inducement and intimidation.[19] In each of these cases, therefore, a medium of communication is a context that, in embedding actions and linking them into systems, constitutes an essential dimension of human coexistence.

It is unclear, however, whether media of communication are distinct from the mental and cognitive conditions and reciprocal actions with which the compositionists analyze sociality, and thus whether they genuinely represent an additional dimension of human coexistence. In the first place, actions as well as mental and cognitive conditions do not presuppose media of communication as they do practices; or, at least, the above theorists do not argue that they do. As a result, media of communication, unlike practices, do not enjoy that sort of distinctness from life conditions that would follow from such conditions systematically presupposing them. Furthermore, systems-theoretical media in the work of Parsons, Habermas, and early Luhmann arguably reduce to configurations of actions and psychological conditions (in conjunction with features of and connections among settings). What I mean is that it is not obviously wrong to think of these media as general ways in which the psyches (and settings) of diverse individuals are so organized and linked that transmission of either information or traces that someone has performed a particular action leads others to carry out certain actions (cf. the fuller formulation in the final section in this chapter). Power, for instance, might be thought of as an organization of psyches within interconnected settings of action which is such that when someone with a particular identity, say, boss, gives an order, and this order is transmitted over physical lines of communication (e.g., telephones, public address systems, fax) to employees, the latter are led to perform certain actions. An analysis of this general form can, I believe, be made of power, money, value commitments, influence (and love and art), though money introduces additional nuances arising from its character as both physical object and representation.

Two points should be made about this reconstruction of media of communication via minds/actions and settings/objects. First, it is a little misleading to call money, power, value commitments, and influence "media of communication," even "symbolic" media. Intuitively, a medium of communication is something over which information flows. As the reconstruction suggests, however, power et al. are psyches, settings, and actions being so organized that the flow of information and traces leads to the performance of certain actions. The real "media" of communication are, first, perception and action (including speech acts), and, second, physical substrates and connections such as sound, telephone lines, satellite transmissions, and puffs of smoke. Psyches can be so organized that, when information and traces are conveyed to people through these media, they are motivated to undertake particular activities. The systems theoretical conceptions of media, then, name salient general configurations of the organization of mind/action/setting, on the basis of which the transmission of information influences motivation and action. I believe that the intuition, that money, power, and the like are a distinct dimension of sociality because they embed individual lives contextually, lives partly from this mischaracterization of general configurations of mind/action/setting as "media."

Second, the necessity of invoking setting, along with mind/action, in the reconstruction of symbolic media of communication, points toward a further error in the compositionist attempt to construe human coexistence as transpiring through mind/action alone. These theorists were not oblivious to the role that organized settings of action play in human sociality (e.g., Tönnies's references to common dwelling spaces, and to physical locality as underlying the commonalities constituting community).[20] But mention of settings, just like reference to such embedding contexts as language and custom, was generally absent from their theoretical depictions of the "substance" of coexistence.

The sweeping assertion that the compositionists neglected enveloping contexts has one important exception. In *Human Encounters in the Social World*, Aron Gurwitsch opened the compositionist tradition to the constitutive significance of history and tradition, to which he had been made sensitive by Heidegger and Count Yorck von Wartenburg. Following analyses of H. Schmalenbach, Gurwitsch distinguished three modes of human being-together: partnership, or being-together in the concern for a common affair (cf. Tönnies's *Gesellschaft*); fusion, or being-together in feelings of being united as "one" in spirit; and membership, or being-together in a life-order (*Lebensordnung;* cf. Tönnies's *Gemeinschaft*). Being-together in a life-order (communalization) is shared embedment in language, tradition, and customs; it is carrying out in common the life that *is* those tradition, customs, and language.[21] In a community, therefore, the hanging-together of lives through mind and action is inherently tied to the historical context formed by tradition, customs, and language. In contrast to partnership and fusion, consequently, membership embraces not only reciprocal actions and mental commonalities, but also a constituting context that carries these commonalities and reciprocities. Being-together in a community is essentially coexisting *within* language, custom, and tradition:

> The particular persons in their relations to one another are to be conceived here only in terms of the whole which is alive in them ... the concrete situations in which they encounter each other are to be conceived only in terms of the life-context that embraces them and, in a profounder sense, they are to be conceived only in terms of its historicality.[22]

Differentiating Gurwitsch's account from those of other compositionists is the constitutive import of language, custom, and tradition for coexistence. As suggested, Tönnies et al. would not deny that reciprocal actions and commonalities in sentiments, wills, and feelings depend at least in some cases on the customs, traditions, and languages in which people are immersed. Gurwitsch alone, however, treats these wider phenomena as the omnipresent *constitutive* context of the hanging-together of lives via mind/action that composes communities. Only he, therefore, incorporates them explicitly into the definition of communities. For other compositionists, the dependency of mind/action upon custom, tradition, and language is contingent and occasional. It does not, consequently, feature in their theoretical formulations of either forms of sociality or types of social formation.

Embedment within tradition, language, and customs bears great resemblance to participation in practices. Each is an immersion within a wider historical context that helps constitute the minds and actions of the individuals concerned. Participation in practices and immersion in language and tradition in fact amount to the same if tradition is understood as the continuation of past practices into the present (the continuing presence of the same practices).

As noted, however, Gurwitsch treats shared immersion in tradition, language, and custom as only one of three forms of human coexistence. The other two, partnership and fusion, are not essentially dependent on enveloping context. Partnership is the free association of people in their fidelity to a common affair, whereas fusion is the joining of people through a shared feeling of unity. Both types of coexistence are a hanging-together of lives that transpires through the minds and actions of participating individuals *simpliciter*. At one point in a footnote Gurwitsch remarks that community is "representative of an ontic relationship at large among human beings."[23] As far as I can see, however, he never develops this thought. He thus overlooks the systematic constitutive significance of tradition, language, and customs (practices). All forms of human coexistence, not just community, involve a hanging-together of lives within practices. Partnerships, for example, are always established and consummated within specific practices (e.g., those of business, recreation, and sex), and are thus at least partly determined by the organization of these practices (cf. Chapter 4, on the articulation of intelligibility). Partnerships are what they are, in other words, only within the relevant practices. Theoretical analyses of this form of coexistence, as a result, must refer generally to the latter.

I note that, although Gurwitsch praises Count Yorck for the "most pregnant expression" of the essential historicality of human life, his circumscription of the constitutive reach of custom and tradition to communities alone suggests that he did not fully appreciate the radicalness of Yorck's ideas (or of their reformulation in Heidegger's *Being and Time*, to which Gurwitsch also refers). Historicality is essential in human life only if *all* concrete modes of human existence are historical, meaning either that these modes *are* history (Heidegger) or that their character inherently derives at least in part from whatever is taken to constitute history. Yorck articulates this second position:

Behavior and historicality are like breathing and atmospherical pressure. . . . a self-examination directed not to an abstract ego but instead to the fullness of myself, finds me historically determined, just as physics cognizes me as cosmically determined. I am history as much as Nature.[24]

In the current book, historicality à la Yorck and Heidegger becomes the thesis that conditions of life and relations among people presuppose practices. Human life is essentially historical in that mind/action exists and people coexist only within and against the background of practices. The implications of this thesis for the nature of history cannot be pursued in the current context.

Mind/Activity, Practices, and Sociality

Compositionists see human coexistence transpiring through mind and action *simpliciter*. The previous section averred that practices constitute a further and essential dimension of sociality. Grounding this thesis is the claim that action/mind implicates practices: those that carry understandings of life conditions, nurture bodily abilities, and/or are carried out at a given moment (see Chapter 4). It follows that the compositional media of coexistence – mental commonalities and reciprocal actions – implicate both practices and the necessity that compositionally coexisting individuals participate in and encounter them. These media of coexistence thus imply two further dimensions of coexistence: (1) the wider socialities opened in the implicated practices, and (2) shared or interconnected participation in these practices on the part of compositionally coexisting individuals. Sociality, consequently, does not amount simply to the existence of groups. It also encompasses immersion in practices, in history à la Yorck and Heidegger.

The following section examines the wider socialities opened in practices. As we shall see, a participant in a practice coexists with diverse people in a variety of ways. Not only are the individuals involved more numerous than those with whom he or she coexists via group membership, but the compositional media of communication are only two of many forms of coexistence established within practices. Before considering this, let me briefly explicate the second sort of practice coexistence just mentioned, the common or interconnected participation in particular practices.

In the section on dispersed practices in Chapter 4, I argued that an act of X-ing presupposes the dispersed practice of X-ing (or an integrative practice that establishes the activity of X-ing). An act of X-ing in fact carries on the practice or activity of X-ing. Reciprocal actions, accordingly, presuppose and carry on one or more practices. Whenever, consequently, people's lives hang together through reciprocal actions, they also hang together through either common participation in the same practice(s) or interconnected participation in different ones. The coexistence involved in waving hello in response to a friend's salutation in the street, for instance, consists not only in reciprocal actions of address, but also in common participation in practices of salutation and friendship. Similarly, the sociality implicated in waving hello to an anonymous bicyclist who signals a turn by tossing his arm in the air consists not merely in one person responding to another, but also in one person's participation in salutation and friendship practices bumping up against another's participation in signaling and bicycling ones. In this way, all coexistence transpiring as reciprocal actions implicates an additional dimension of sociality consisting in either participation in the same or highly similar[25] practices and/or ordered participation in different ones.

In Chapter 4, I also indicated that which conditions of life (including actions) are expressed by behavior and inner episodes depends on both the particular practices a person is carrying on at the time of the behaviors/episodes concerned and a nexus of practices that carries understandings of life condition concepts. Since expressing and being in particular mental and cognitive conditions presuppose practices as well as participation in (and encounters with) them, any coexistence composed by an organization of different people's psychological conditions likewise implicates a further dimension of sociality consisting in participation in the same practices along with ordered participation in different ones. In particular, commonality in mental and cognitive conditions entails participation in and exposure to some assortment of the same or highly similar practices.

Recall, in addition, my earlier contention, that to be an individual is most centrally to be in conditions of life. Since having mind and performing actions implicate practices as well as participation in them, so, too, does the existence of an individual. An individual's existence also thereby entails immersion in (1) the socialities opened in those implicated practices in which he or she participates, and (2) socialities consisting in common or ordered participation in the same, highly similar, or different practices. In other words, being an individual and existing socially are coordinate and simultaneous achievements in human life: any individual subject is ipso facto a social being. The inherent "socialization" of individuals that comes to pass within practices indicates once again that any approach to coexistence that sets out from the individual presupposes practices, contrary to its intention, as a further site of sociality.

The Sociality Established in Practices

Any practice opens a dense field of coexistence embracing its participants. This section surveys several forms of sociality opened in practices that significantly impact life there. Illustrating an earlier declaration that the present account of sociality is incomplete, my discussion neither considers all forms of coexistence established in practices nor exhausts the examined forms.

A first form of coexistence is the interpersonal structuring of understandings, rules, and teleoaffective orders. This dimension of sociality has two basic forms: commonality and orchestration. "Commonality" means that the same understanding, rule, or component of teleoaffectivity (end, task, emotion, etc.) governs different people's behavior. Vis-à-vis dispersed practices, commonality more or less pertains to understanding alone: Participants in any such given practice act out of the same understanding. Vis-à-vis integrative practices, commonality typically embraces all three types of action-governing factor: A participant's behavior is governed by understandings, rules, and mental as well as cognitive conditions that also govern the behavior of some, though not all, other participants. Commonalities in these factors do not embrace all participants in an integrative practice because participants not only act out of or observe different combinations of its organizing understandings and rules, but differentially instantiate its normative field of teleoaffective orders. Any integrative practice exhibits, as a result, a range of commonalities in conditions and rules that embrace different, overlapping sets of participants.

Participants in an integrative practice also face together various subsectors of its teleoaffective structure. This fact further explodes the compositionist viewpoint. Unlike commonalities in understandings, rules, and mental/cognitive conditions, a field of possibility is not a candidate for reduction to features of individuals. It is emphatically the property of a practice.

Although social theory has long acknowledged commonalities in action-governing phenomena as a central form of sociality, it has paid considerably less attention to the orchestration of such phenomena within integrative practices. (Orchestration is not found in dispersed practices.) "Orchestration" consists in different understandings, rules, and conditions nonindependently determining what different people do. It exists either when a particular factor governs one's person's behavior because some other factor governs other people's behavior or when some third thing determines the nonindependence. An example of the first sort of orchestration is one person's pursuit of a particular goal in business practices reflecting other participants' pursuit of different goals; or one participant's caring about business affairs in a certain way depending on the ends a second participant pursues. An important subtype of the second sort of orchestration is the deliberate establishment of orchestration. Examples are sundry people observing different rules in military practices and the assignment of tasks to military personnel of one rank hanging together with personnel of other ranks possessing particular knowledges. Of course, like commonalities, much orchestration is neither planned, intended, or even apprehended by those encompassed. Although, for example, bosses, agents, secretaries, and the like interdependently care about business matters differently, in most cases this fact is neither planned nor set up by any of them (or anyone else), and in some cases goes unnoticed.

Social theory's one-sided focus on commonality at the expense of orchestration marks its treatment of all three sorts of action-governing factor. The idea, for instance, that the glue of social life is observance of the same rules pervades sociology – for example, the Durkheim-Parsons-Habermas notion of a normative consensus, Weber's concept of an order (a uniformity in social action brought about by orientation to particular "specifiable [*angebbare*] maxims"), and George Homans's archetypical construal of institutions as rule-governed regularities in behavior.[26] Theorists have of course acknowledged that different actors interdependently observe different rules. But recognition of orchestration has primarily featured in analyses of institutions and roles. For instance, Peter Berger and Thomas Luckmann write that one component of the knowledge distributed among roles within an institution is the "recipe knowledge" (Schütz's expression) that supplies appropriate rules of conduct for any particular role.[27] Since the distribution of knowledge among roles is orchestrated (so that roles mesh), so, too, is the resulting distribution of rules observed by different people. Orchestration, however, occurs far more prevalently than in roles and institutions alone. As a fundamental dimension of the sociality established in all integrative practices, it is an omnipresent feature of social life.

The sociality of any integrative practice embraces a rich interpersonal structuring of that which governs action. Within this structuring, participants' lives hang together through sameness and difference. The understandings, rules, and conditions that govern one participant's behavior both overlap with those and hang together with the different factors that govern others' behavior. As a result, a participant in an integrative practice, through the governance of action, coexists in multifarious ways with different combinations of other participants.

A second general form of sociality within practices is one person being the object of another's life conditions (including understandings and actions). Two lives hang together, for instance, when one person knows or thinks something about someone else, when the second person is the object of an emotion, feeling, or action of the first, when the first understands or intends to do something with regard to the second, and so on. That coexistence within a practice embraces participants being the objects of one another's conditions is an obvious fact requiring little comment. Within teaching practices, for instance, the lives of teachers and students hang together partially through teachers acting toward students, having opinions about their abilities, understanding their behavior, setting goals for them, caring about them, and so on. Being directed toward others is in fact such an omnipresent and salient feature of everyday life that individualist thinking has occasionally christened it the central locus of sociality. Weber, for instance, famously defined as social any action that is oriented by an actor's directedness toward the behavior of others. On this definition, action is social when it is governed by an intention, belief, hope, and/or expectation concerning others' behavior. Combined with Weber's subsequent stipulation that a social relation consists in the chance of social action occurring, this definition treats sociality as above all the actual and possible orientation of actors to one another. Types of social formation are then differentiated through different aspects and possibilities of such orientation.

Earlier I wrote that lives hang together in ways having nothing to do with people's encounters with others. People likewise coexist apart from their orientations to one another. Schütz's and Weber's restriction of sociality to actors' directness toward others overlooks these additional media of coexistence, for instance, the commonalities and orchestrations in action-governing factors through which lives hang together without being focused on one another. Orientations toward and encounters with others are perhaps the most experientially significant and obvious forms of sociality. But anyone focused on others coexists with these and further others in a multitude of ways.

A third general medium of sociality is settings of action and the spaces established there. Socialities in this medium divide into those embracing single settings (commonality) and those encompassing multiple ones (orchestration). The lives of participants in a practice often hang together by virtue of taking place within the same or different settings.

An important type of commonality is the simultaneous presence of different participants in a single setting. This state of affairs often, though not always (especially when nonparticipants or many participants are present), occasions face-to-face interaction. A second type of single-setting sociality consists in a particular event or state of affairs initiating a new situation for a group of participants. The arrival of a teacher in a classroom, for example, ushers in a new situation for at least most of the schoolchildren present. A third type of commonality concerns the layout of objects in a given setting and the spaces thereby set up. In Chapter 4, I discussed how practices open spaces of places at which their constituent doings and sayings are correctly and acceptably performed. These places are anchored in objects, which are combined into settings. A setting is thus a particular (experientially circumscribed) configuration of objects that anchors a space of places. A classroom, for instance, is an organized nexus of "equipment" for sitting, standing, writing, painting, reading together, and the like that anchors places where the actions performed in teaching, reading, writing, painting, and so on can correctly and acceptably take place. As discussed, moreover, settings are often set up as sites where a given practice is to take place. Classrooms, for instance, are laid out to be settings where the actions of educational practices take place, whereas a factory assembly room is set up to be a setting where actions of industrial activity occur. When, consequently, participants in either practice act within the respective setting, their lives hang together via the setup of objects in that setting and the common space of places opened there by the organization of the practice involved. Remember that in these settings people can carry on practices different than those for which the settings were set up. When this occurs, it is much less likely that their lives there hang together through a common space of places.

A fourth type of commonality in settings (there are more) consists in different participants nonindependently performing different actions at different times in a given setting. An important subtype embraces "switching settings." A switching setting is a locale where different participants perform coordinated actions at different times because of some feature of it. A particular street corner might be a switching setting, for instance, because the presence of a mailbox leads different people to deposit and collect mail there. Switching settings are important because they make certain accomplishments possible. The transfer of mail from sender to recipient, for instance, is an accomplishment effected by a set of coordinated actions performed at different times in a series of switching settings: street corners, post office sorting rooms, truck loading docks, airport mail rooms, and so on.

People's lives hang together by virtue of transpiring in not only the same, but also different settings. Participants in a given practice coexist in this general manner through three sorts of multiple setting features: nonindependent situation-initiating entities, intersetting layouts of objects and spaces, and physical connections. The first type of orchestration consists in different events in different settings nonindependently constituting new situations for participants in those settings. An example is provided by a bank robbery, in which different but interrelated signals are given in different settings with the purpose of bringing about a set of coordinated actions by a particular group of people engaged in criminal practices. An example in which the different initiating events are type identical is a special report interrupting normal broadcasting and bringing about new situations for those gathered about their television sets.

The second form of multiple-setting sociality is intersetting layouts and spatialities. An intersetting layout is a setup of different settings, together with the physical connections among them, that coordinates and facilitates the performance in those settings of different constituent actions of a given practice. An intersetting spatiality is the correlated field of intermeshing normative spatialities that is anchored in the different settings encompassed in an intersetting layout. Intersetting layouts and spatialities exist wherever objects that trigger, call out, or signal particular ways of acting are distributed in a nonindependent fashion among settings. An example is a layout of stop signs, crosswalks, traffic signals, and other traffic control equipment. When participants in an integrative practice act in settings that are embraced by an intersetting layout/spatiality established in the practice, their lives hang together via this structure. Intersetting structures of this sort are found in the organization of a factory, the design of a new highway system, and in the construction of a submarine. Note that when organized settings are physically discontiguous, they are linked by physical or technological media of access or communication (road, telephone lines, etc.).

The third sort of orchestrated setting sociality, physical connections among settings, comes in three basic forms: continuous physical structures, avenues of access, and technological connections. Continuous physical structures are exactly that – natural and man-made structures that stretch between settings. Examples are rivers, bridges, buildings, and mesa tops. Avenues of access are physical passageways by which access to and exit from settings are effected (e.g., mountain passes, rivers, doorways, bridges, open spaces, and hallways). Technological connections, finally, are media of communication. Examples are telephone systems, satellite communication systems, and fiber optic cables. Technological connections are especially important because they link physically nonadjacent settings, thus promoting the meshing of action over objective spatial distance. Notice that physical connections among settings are not always keyed to particular practices. Constructed connections such as bridges are not usually set up solely to link participants in some particular practice(s); natural physical connections such as rivers often relate participants in different practices; and technologically linked people sometimes participate in different practices while linked.

The final form of sociality within a practice to be discussed in this section is chains of action. As noted in Chapter 4, a chain of action is a series of actions, each member of which (apart from the first) is a response either to the immediately preceding member or to an event or change in the world directly brought about by the preceding member. An example is a general issuing an order, which is delivered by an aide to the radio dispatcher, who in turn reads it over the shortwave, upon reception of which the division commanders give their own orders, and so on. The following features of chains of actions should be noted: (1) A multiplicity of action chains can branch out from a given origin; (2) successive actions in a given chain can be performed in the same setting at the same time (e.g., face-to-face interaction), in the same setting at different times (e.g., in switching settings), in different settings at the same time (e.g., non–face-to-face interaction), or in different settings at different times; (3) physical connections between settings support chains of action; and (4) most important, each link in a chain is mediated by understandings, rules, and/or teleoaffective orders, since these govern what someone does in response to a previous action or to the changes the latter directly brings about. Note that two people's lives hang together not only when actions of theirs are part of a single chain of action, but also when action(s) that each performs individually are elements of different chains with the same origin.

Human lives hang together via an immense, essentially unsurveyable labyrinth of action chains that branch, cross, coalesce, terminate, and begin new. Some chains percolate within particular practices, whereas others connect practices and help weave them into nexuses. The foregoing paragraph described a chain of action enclosed within military practices. Usually, however, not all chains that a person is entangled in when participating in a given practice are confined to the practice involved. He or she links onto chains originating in different practices, just as participants in other practices link onto chains that he or she perpetuates, therewith extending these chains into those practices. The following example illustrates these possibilities: A rancher places a no trespassing sign on the edge of her property; a hiker who comes upon the sign later that day turns back; the rancher, returning the next day with her husband, spots the footprints and smiles; her husband notices her smile and asks if she's no longer angry. Chains of action are confined within a given practice only to the extent that the organization of the practice governs their links. Despite this restriction, action chains are clearly a significant feature of any practice, dispersed or integrated. In addition to acts of X-ing, for example, the dispersed practice of X-ing embraces reactions to such acts, thus minichains of action. Such integrated practices as those of farming, education, cooking, and business, meanwhile, embrace extensive chains through which the ends pursued in those practices are sought. In participating in a practice, consequently, a person invariably enters myriad chains of action embracing other participants. His or her life comes through causality to hang together with those of others.

In sum, a practice harbors an extensive tissue of sociality encompassing chains of action, directness toward others, physical connections among settings, and commonalities in as well as orchestrations of understandings, rules, teleoaffectivity, and settings. I add that tissues of coexistence embrace not merely those people who participate in a practice at a particular time, but past and future participants as well. I will not, however, consider complexities introduced by the temporal structure of practice sociality.

I wrote in Chapter 1 that practice theorists consider practices the central phenomenon in social life to a great extent because they view them as the site where intelligibility is articulated. It is important, consequently, to explain how the foregoing enumeration of forms of sociality highlights the role of normativized intelligibility in human coexistence. To begin with, commonalities in and orchestrations of rules, understandings, and mental as well as cognitive conditions are commonalities and orchestrations in the structure of action intelligibility. For these three types of factor determine action by structuring the action intelligibility governing it. Established within the normative organizations of practices, the forms of sociality that are such commonalities and orchestrations are configurations of normativized intelligibility.

Directness toward others, furthermore, articulates the significance of those others. Any belief, expectation, hope, fear, or the like concerning other people assigns meaning to them by specifying what they are (for someone): individuals or groups who possess certain traits, from whom so and so can be expected, of whom such and such is to be hoped, who are to be feared, and so on. This articulation of world intelligibility, which like the articulation of action intelligibility is subject to normativizing circumscription through the organizations of practices, in turn structures directness toward others in intentions and actions. The form of coexistence that is people being the objects of one another's conditions is thus a normativized articulation of the world intelligibility pertaining to others.

Most forms of coexistence that transpire through settings and spaces, meanwhile, are mediated by normativized intelligibility. The articulation of intra- and intersetting spaces is a component of world intelligibility, just as are the significances of intra- and intersetting situation-initiating events, of the presence of others, of orchestrated layouts of objects across settings, and of the objects around which switching settings coalesce. Like the meanings of others, those of objects and settings in turn structure the action intelligibility that governs performances in those settings. Only physical connections among settings retain a form of independence from world intelligibility. Physical connections are distinct, self-existent phenomena regardless of whether they or any part of them are ensnared within the intelligibility articulated in practices. They, along with objects, constitute a material structure through which people move, one that, unlike other media of coexistence, would remain even if all human beings suddenly disappeared.

As noted, finally, intelligibility also underlies chains of action. In particular, interpersonal structuring of intelligibility mediates links in such chains.

Hence, the sociality established in a practice is centrally spelled out within and mediated by the intelligibility articulated in that practice. Human coexistence within a practice consists most crucially, though not exclusively, in samenesses and nonindependent differences in action and world intelligibilities through which participants live in diverse, physically interconnected settings and are caught up in causal lines of action that traverse these settings.

This claim that sociality is above all played out within and mediated by intelligibility, is not open to the sort of charge that Habermas once directed at Gadamer's hermeneutics, namely, the propagation of a linguistic idealism that, in confining all aspects of social life within linguistically articulated intelligibility, is unable to grasp adequately such social phenomena as work, domination, and ideology that act upon and distort meaning.[28] I think that Gadamer was right to reply that work, domination, and ideology are not, as Habermas seemingly portrayed them, independent from but are instead mediated by world intelligibility.[29] But Gadamer cramped his position and invited Habermas's criticisms by maintaining that intelligibility, the medium of being-in-the-world, is articulated in language. By treating language as the site of the coming-into-being of world, he problematized it *qua* something within the world and cloaked it with an updated, historicized version of the transcendental garb with which Kant had adorned the faculties of understanding and sensibility. In my account, by contrast, intelligibility is articulated firstly in practices, not language. And practices, in embracing not only articulations of intelligibility, but also actions and chains of actions along with setups of and connections among settings and spaces, are inherently worldly entities. This also holds for the tissues of coexistence that practices establish, which similarly weave together structurings of intelligibility with chains of action and setups of settings and spaces. Furthermore, since human life transpires within practices, practices are the site where work is performed, domination effected, and ideology propagated. All in all, a social ontology that adopts the meshing of intelligibility, world, and causality within practices as its ontological point of departure in no way compromises the "reality" of these key social phenomena. The thesis that sociality transpires through intelligibility, far from prosecuting a meaning idealism, integrates intelligibility into the founding institution of human coexistence where its constitutive and mediative role not only in work, domination, and ideology, but social affairs generally, can be approached and theorized.

I conclude this section by pointing out that the tissues of coexistence established in integrative practices undercut the contrast between *Gemeinschaft* and *Gesellschaft* as fundamental categorization of types of sociality. The compositionists spelled out this, their master contrast in terms of two further oppositions, those between unintentional and intentional and nonset-up and set-up. Whereas the sociality of community and communalization is an unintentional and nonset-up commonality in mental and cognitive conditions, that of society and societization is an intentional, set-up coordination of wills and behaviors. Undermining the elementariness of this distinction is the fact that the coexistence established in integrative practices is both partly and partly not set-up. Any such practice, for example, exhibits a mix of unintended and intended commonalities in and orchestrations of understandings, rules, and teleoaffectivities. Similarly, the chains of actions propagating within it are partly undesigned and partly fashioned. The tissue of sociality opened in a practice thus embraces at once the unreflective, automatic enveloping commonalities and the reflective, implemented limited arrangements immemorialized in the distinction between community and society.[30]

Social Order Within Practices

All states of sociality, I wrote earlier, embrace a social order(ing). Whereas sociality is the hanging-together of human lives, social order is the arrangement of lives that characterizes a *Zusammenhang* of them. Since a practice opens a tissue of coexistences enveloping its participants, it also automatically establishes orderings among them.

An important aspect of social order is the different locations individuals occupy in the arrangement of lives. A person's location within a practice's social ordering is his location vis-à-vis others as established by his place in the tissue of coexistences opened in the practice – thus by commonalities and orchestrations in the intelligibility governing his and their behavior, how others are the objects of his conditions and he the object of theirs, the connections of his life with theirs through settings, and the particular chains of actions that link him with others. This phenomenon of differential locatedness in webs of coexistence is what in the domain of practices corresponds to the widespread and familiar notion of differential position in social structure. Although I will not discuss social structure in this chapter, I claim ex cathedra that differential position in "social structure" is grounded in differential locations in practice socialities.

The social ordering established in a practice contains multiple places because of the uneven and asymmetrical "distribution" of participants among the commonalities, orchestrations, chains, and directednesses that constitute the practice's sociality. Participants, for instance, are party to different subsets of the commonalities found in the practice. Further differentiating them are the different directednesses, chains, and orchestrations of intelligibility and settings in which they are embroiled. These directednesses, chains, and orchestrations are also asymmetrical: A is directed toward B differently than B toward A; a particular stretch of a chain of actions flows from A to B but not vice versa; the entities A deals with in one setting differ from those connected ones B deals with in another; and so on. Participants in a practice are clearly not equal within the webs of coexistence opened there. They are instead separated, hierarchized, and distributed. This is a crucial feature of existence within a practice that was passed over in my initial discussion in Chapter 4.

It is not possible in the current context to offer an account of social ordering and the differential locatedness characterizing it. I aim simply to convey a feeling for the character of these phenomena, and will attempt this by sketching the role that social position plays in the constitution of differentiality. I hope that this sketch will also suggest how an analysis of social order as arrangements of lives contributes to an analysis of social order in the more familiar sense of regular and regulated nonovertly violent human coexistence.

I note first that the notion of a social position descends from that of a role. For many decades in the twentieth century, social theorists of various stripes treated roles as the point of connection between individuals and society, in Parsons's words, as "the primary point of direct articulation between the personality of the individual and the structure of the social system."[31] Although roles served in theoretical analyses as axes about which arrangements of lives were organized, above all in institutions (via the interlocking patterns of behavioral expectations and prescriptions that constitute roles), recent criticisms of the notion of roles problematize its usefulness in the analysis of coexistence.[32] The notion of a social position overcomes these criticisms, for instance, that role analysis treats the behavioral expectations and prescriptions that constitute roles as fixed, inalterable, and the object of consensus. This descendant notion thus offers itself as a more propitious tool with which to analyze differential locatedness in practice coexistence.

In Chapter 1, I described Chantal Mouffe's idea that a person's identity is an ensemble of subject positions woven around core determinations called "nodal points." Mouffe is not completely clear about what a "subject position" or "determination" is. I will interpret the idea as follows: A subject position is a signifier, or linguistic expression, that categorizes people and receives its classificatory meaning from its use in particular practices. The practices involved are sometimes few in number, as with the subject positions "baker" and "professor," and sometimes extensive, as with "African American" and "Norwegian." The meaning, moreover, of a subject position can be multiple and fluid. It can vary among the practices in which the determination is wielded; and also among the individuals who wield it, depending on, for example, the combination of practices in which they encounter and employ the term and, within those practices, the particular contexts in which and the particular individuals to whom it is applied. In most cases, individual idiosyncracies concerning a specific position play off the normativized and regularized, though potentially plural meanings established in specific practices. A position's meanings can also evolve along with changes in practices. In any event, a person's identity consists in the collection of subject positions she assumes in participating in a range of practices.

Sociality within practices is organized to varying extents around the subject positions made available in those practices. That is to say, the coexistences opened in a practice often are spelled out in terms of or associated with those positions. Assuming and being identified with a particular position operative in a practice assigns meaning to a person for all participants, including herself. When someone is understood to be a baker, professor, African American, or Norwegian, a range of beliefs, expectations, hopes, feelings, understandings, and the like (varying among participants within bounds and amidst commonality) apply to that person. This intelligibility helps determine how others are behaviorally directed toward her and how she, in turn, is directed toward them. It also helps determine how she perpetuates or initiates chains of action when she knows to whom she is (re)acting (e.g., a baker, a Norwegian). Orchestrations of intelligibility are also organized around subject positions and for much the same reasons. That, for instance, one participant's pursuit of a given goal in business practices depends on a second participant's pursuit of a different goal often reflects the positions of the individuals involved (e.g., boss and courier). Rules, furthermore, regularly assign different actions, duties, or forbearances to the occupants of different positions, just as teleoaffective structures standardly evince a patterning of ends, projects, tasks, and even emotions among positions. The divisions of labor in, say, farming, industrial, and business practices exhibit teleological distributions of this sort.

Moreover, sociality is not only laid out by reference to but also comes to be contingently associated with subject positions. For instance, specific chains of action and orchestrations of intelligibility might become associated in a practice with a particular position and thereby transform its intelligibility. "Boss" might become a more admirable position, for example, if participants in business practices develop warm, sympathetic feelings for bosses in reaction to kind, employee-fostering actions of particular bosses.

Of course, the people with whom one coexists in a given practice have identities composed of various subject positions, only some of which might be organizational foci in the practice. Although the organization of a practice – along with the action chains and layouts of as well as connections among settings coordinate with and following from this organization – revolves partly around specific subject positions, the individuals occupying those positions often have complex and unstable identities cobbled together through their participation in myriad practices. To the extent that people's behavior reflects the full complexity of their identities, the unfolding of interaction and coexistence within a particular practice escapes and is potentially at odds with the practice's organization. This situation raises familiar questions about the channeling of interactions within practices (and institutions and social formations more broadly), which cannot be addressed here.

The sociality opened in a practice is nonetheless pivotally organized around a range of subject positions. For commonalities in and orchestrations of intelligibility and settings, as well as asymmetric directednesses and chains of action, are organized to varying extents around such positions. So the arrangement of lives established within a practice also rotates around the subject positions that circulate there; and where any individual finds herself within this arrangement depends crucially on her identity. Differential locatedness in social order, consequently, rests squarely on people's subject positions. Analyzing social order thus requires ascertaining the identities offered and adopted within practices.

The Constitution of the Social Field

In retreating before the task of more fully analyzing social order, I have begun to chart the limits of the present discussion. Until now I have been considering the sociality opened for participants in particular practices. As just came to the fore, however, people obviously participate in multiple practices. Besides implying that they are enmeshed in a variety of tissues of sociality, each associated with a given practice, this commonplace also entails that coexistence extends across practices. The conception of the social field as a heterogenous and modulating weave of practices likewise implicates this wider sociality, for interconnections among practices ensure that individuals coexist not only with those who participate in the same practices as they, but also those who participate in practices connected with theirs. The lives of a Swedish financier and baker hang together, for instance, through connections between Swedish banking and baking practices (e.g., the use of the products of one in receptions in the other), just as the lives of the Swedish financier and a Singaporean baker hang together through connections between Swedish banking practices and Singaporean baking ones (e.g., the use of Singaporean sweets in the receptions). That the lives of participants in different practices hang together through connections among those practices shows that the preceding discussion of sociality is just part one of a more complete ontological analysis of the social field. The following remarks do not provide this analysis, but instead sketch general tasks and features of the practice theoretical social ontology opened up by the current chapter.

One general task for a wider account of the social field is analyzing the nature of social order. A second is identifying the relations among practices by virtue of which they form nexuses. I believe that the forms of sociality through which the lives of participants in a given practice hang together also represent central types of relation among practices. The reason for this coincidence, intuitively speaking, is that to say that two practices are connected is to say that life in one hangs together with life in another. Relations among them, as a result, transpire through and as the hanging-together of the lives of their respective participants. (These relations also transpire through and as the very constitution of the lives of those people who participate in both.) Now, lives in different practices hang together through the same sorts of state of affairs as lives in a given practice do: commonalities in and orchestrations of intelligibility and setting, directedness toward others, connections among settings, and chains of actions. Which of these types of relation are more prominent interpractice media of coexistence varies among different types of social formation (e.g., economy, family, artistic movement; see ensuing discussion). Although I will not spell this out here, this account of the relations among practices implies that the only principles that pertain to the sociality established in nexuses of practices beyond those pertaining to the sociality opened in individual practices are principles introduced by the change of *scope* effected in focusing on multiple practices instead of individual ones, for example, principles pertaining to wider spatial breadth and increasing numbers of people. This claim resembles Randall Collins's thesis, that all differences between micro- and macro-accounts of social life arise solely from the increases in space, time, and number that characterize the wider swaths of social life investigated in macro inquiries.[33]

A third important task facing a practice theoretical social ontology is analyzing the constitution of social formations. By the "constitution" of a social formation I mean what the formation consists in, of what it is composed. A formation's constitution is not the same as its definition. The expression "social formation," moreover, is used in an extremely general sense to designate any consolidation of coexistence embracing the lives of two or more persons. This expression corresponds in its generality to the expression *soziale Gebilde* as used by most German social theorists before (and many after) World War II. Most of the familiar entities that social scientists study and catalog are social formations, from conversations, fights, and interactions in general to families, clubs, parties, and artistic movements, from armies, governments, trade associations, and economies to social unrest, revolutions, and the development of new technologies. Types of social formations include exchange networks, associations, movements, groups, institutions, historical events, and systems (in a loose sense).

These types of entity of course have disparate constitutions. But the theses that the social field is a nexus of practice, and that sociality therefore consists fundamentally in the coexistences opened in nexuses of practice, entails a general principle, namely, that the constitution of any social phenomena be understood, in one way or another, by reference to such nexuses. I understand this principle to imply that most social formations can be understood as consisting in slices, or subconfigurations, of the total labyrinth of interconnecting practices that constitutes the social field. A social formation consists in a particular intermeshing of specific practices that encompasses specific, sometimes open sets of individuals – it is a particular "bundling" of practices. The sociality that composes it is thus the multidimensional and multilayered lattice of coexistences embraced by that bundling.

A more pictorial elucidation of this conception builds on a metaphor found in Wittgenstein. Wittgenstein occasionally speaks of life as a weave, or rug (*Teppich;* e.g., *PI,* p. 174b). Imagine a solid three-dimensional fabric in the form of a cylinder that is woven out of immensely many fibers of various colors and lengths that interweave in extremely intricate ways. In this image, the cylinder corresponds to the social field over space and time and fibers of a given color to the carrying-on of a given integrative or dispersed practice at particular places for particular lengths of time. (Keep in mind that whenever, as is usual, an integrative practice varies over space and time, the carrying-on of it at a given locale for a particular time is the carrying-on of a particular version of it.) A social formation consists in a large or small, possibly irregular, convoluted, or discontinuous chunk of the cylinder: a particular interwoven mass of fibers. Moreover, the "bundling" of practices that composes a given social formation corresponds to the specific interwovenness of particular fibers that composes a given chunk. (The chunks, be it noted, are not mutually exclusive, i.e., formations overlap.) So a social formation consists in a particular intermeshing of particular practices that encompasses a specific, sometimes open set of individuals; and, as the image suggests, the practices concerned intermesh per states of affairs (see subsequent examples) that have to do with, and for the most part arise from, the carrying-on of the practices involved at certain times and places.

I wrote earlier that, intuitively, two practices are interconnected when life in one hangs together with life in another. This also holds for the bundling of practices that composes a social formation. Any such bundling embraces a specific concatenation of ways in which life hangs together across certain practices. This is the same as saying that any such bundling embraces a constellation of ways in which a particular, possibly open-ended collection of lives hangs together across these practices. Now, as just suggested, practices bundle through states of affairs that largely come to pass through the carrying-on of the practices involved at particular times and places. It follows that the states of affairs through which lives hang together across certain practices – and social formations come to be and persist – largely exist through these lives carrying on those practices. Different mixes of such states of affair characterize different types of social formation. A family, for instance, consists in a specific bundling together of (often local versions of) such practices as those of sleeping, cooking, rearing, recreation, and hygiene that embrace specific, usually biologically related individuals. The total state of coexistence encompassed by a given family is all the ways that family members' lives hang together as that intermeshing of practices – in other words, all the ways their lives hang together across these practices through states of affairs established by their carrying-on the practices involved. These states of affairs include: (1) the same biological individuals participating in the different practices; (2) the spatial consanguinity of the settings in which they do so (e.g., living quarters); (3) particular actions (e.g., slicing vegetables) constituting moves in more than one such practice (e.g., cooking and disciplinary practices); (4) particular causal links among actions in them; and (5) commonalities in and orchestrations of intelligibility organized around the social positions occupied by family members in the bundled practices (e.g., "mother" and "daughter," "cook" and "gardener," "momma's boy" and "lazy bum").

An economy is likewise a social formation and subject consequently to the same general sort of analysis. The (often local versions of) practices bundled in it include those of industry, finance, farming, transportation, management, lobbying, bargaining, buying, advertisement, and, through commercialization, recreation, entertainment, editorship, and celebration. Unlike in a family, the total state of sociality encompassed in the bundling of these practices embraces an open-ended and constantly evolving set of individuals who need not exhibit any particular biological relations and often gain access to various domains of the economy on the strength of specific elements of their identities. Bundling together, moreover, these practices as the economy, that through which life hangs together across these practices, is a plethora of phenomena, for example, (1) particular communications connections among settings; (2) the layouts of pro-

duction sites and department stores; (3) orchestrations and commonalities in intelligibility and settings that constitute hierarchical organizations; (4) convergences in interests; (5) monetary offers; and (6)
elaborate chains of action stretching from corporate headquarters to
government offices to auto repair shops and neighborhood planning
groups. Emerging from this elaborate weave is a multidimensional
ordering of lives within and across the bundled practices. In principle, therefore, the constitution of the economy is no different from
that of a family. It simply embraces wider spatial breadth, more
complex networks of chains of action, larger numbers of individuals
and practices, and a greater range of types of commonality in and
orchestration of intelligibility and settings.

Not only are specific social formations such as families and economies ontologically analyzable in this fashion, but the constitution of
such types of social formation as institutions and groups can be
characterized by reference to possible features of bundled practices.
Beyond social formations, other types of social phenomena such as
power, gender discrimination, and social structure can be analyzed
by reference to other features of the nexus of practices. I will not
defend or fill in these ideas further. I presume that these sketchy
remarks provide some, albeit minimal, feeling for both the third task
posed to a practice theoretical social ontology and the shape such an
ontology would generally take.

Before continuing, it is appropriate to repeat a comment made in
the first section of this chapter, that the present account of sociality
does not presume to capture every aspect of human coexistence.
Conceiving of social formations as consisting in bundlings of practices, specific concatenations of ways in which life hangs together
across particular practices, does not in itself reveal all aspects of the
human togetherness composing them. It instead spells out the basic
constitution of a social formation and implies that further details and
complexities of life and coexistence in one are features of life and
coexistence within a particular bundling of practices – further ways
life proceeds and hangs together within and across these practices.
The current account thus constitutes a framework through which
these formations can be investigated and their intricacies disclosed.

This is also a good juncture to pause and recall that, in highlighting
complexity and difference within the social field, the picture of the
social as a nexus of practices rejects the presence of well-defined
large-scale social unities. Having just made a few comments about
economies, I will illustrate this repudiation by confronting Habermas's version of the idea that the economy is a well-defined system
of action. My thesis, simply put, is that the variety of actions, prac-

tices, and relations among actions or practices that characterizes a real economy undercuts the ontological value of this idea.

Habermas follows Parsons in conceiving of a system as a boundary-maintaining network of actions, held together by a particular medium of communication, that coalesces over feedback out of the totality of actions in fulfillment of a particular function of the totality. For Habermas, a modern economy is a system of success-oriented purposive-rational actions that is coordinated in a nonconsensual manner via money and, through feedback, differentiates out of (and becomes stabilized within) the totality of actions in fulfillment of the function Parsons called "adaption": the maintenance of the material substrate. The "environments" over against which this system maintains its boundaries are private households and public administration. According to Habermas, the economy becomes a system when exchanges between it and its environments are steered by the same medium that regulates its internal transactions.[34] "For a medium-steered subsystem to take shape, it appears to be sufficient that boundaries arise across which a simple interchange, steered by a *single* medium, can take place with every environment."[35]

The oversimplification countenanced in this analysis is painful. The idea that a social entity as massive as a modern economy could be constituted by a single sort of action, success-oriented purposive-rational action, clearly overlooks the variety of emotional, traditional, and even value-rational actions (to use Weber's typology), or normative, dramaturgical, and communicative actions (to use Habermas's), that also help constitute any such large-scale social entity. Habermas has here – surprisingly – fallen into line with an economistic gaze that treats actions of these further types as inessential adornments to what it construes as the central core type in economic life. An oversimplification of this sort may be justifiable for the purpose of model building in economics and other social disciplines, but it is not defendable in social ontology. In addition to discrediting the multifarious types of action that constitute economic life, it forecloses recognition that a modern economy bundles a range of *practices*, from business, industrial, farming, and financial practices, and managerial, conspiratorial, and human management ones, to those of recreation, political decision making, celebration, parenting, the media, and the like. The economy encompasses an elaborate constellation of interconnected practices that is poorly represented simply as a networking of purposive-rational actions over money.

Moreover, it is not possible, even on Habermas's own terms, to delimit precisely the boundaries of the system said to be the economy. The phenomena of success-oriented purposive-rational action and of money as medium of action coordination do not by themselves mark these boundaries. For actions of this sort and the use of money as a means of nonconsensual coordination are also found in the economy's so-called "environments," private households and public administration (as well as other domains of social life). Mere reference to the "function" of maintaining the material substrate does not accomplish this feat either, since, strictly speaking, it is not the economy alone but it in conjunction with private households and public administration (and what else connects with these) that fulfills this function. In this connection, recall recent analyses of the role households play in maintaining the labor force.[36] Furthermore, since Habermas fails to explain exactly how feedback works on his account,[37] he cannot employ the idea that systems coalesce over feedback to mark which actions constitute the economy. Appending the claim that feedback coalescence occurs over money does not rectify this situation, for this addendum does not explicate what it is for feedback to stabilize actions (as a system) over money when their unintentional consequences so mesh as to maintain the material substrate. This can be seen from the fact that money can coordinate success-oriented actions regardless of whether "feedback" occurs. Nor, finally, can the notion of a "systematic interdependency" (*systemischen Zusammenhangs*)[38] among actions mark the boundaries of the economy, since this notion is itself explicated via feedback, stabilization, "adaption," and money.

Additional problems arise upon leaving Habermas's chosen terrain. Not only the economy, but public administration and private households too, consist in bundlings of practices, elaborate concatenations of ways in which life hangs together across certain practices. These various bundlings are obviously so intricately interlaced that the notion, even the self-consciously oversimplified modeling supposition, that life hangs together across them through a single medium of communication, is fantastic. A vast range of directednesses, action chains, connections in settings, and commonalities in as well as orchestrations of intelligibility link life across the bundles of practices composing these formations. The constellations of bundled practices that compose the regions of the social field designated "private household – economy – public administration" are in fact so complex that the further ideas that these regions are stable, that they are stabilized through feedback, and that they have the sole "function" (if one wants to talk this way) of maintaining the material substrate are equally unrealistic.

I emphasize that acknowledging the complexity of the bundlings that constitute such social formations as modern economies does not imply abandoning generalizations about, or all hope of discovering approximating large-scale principles roughly applicable to, these formations. Many generalizations do in fact hold of the economy: The strategic use of money is far more prevalent in the economy than in other domains of life; the economy therefore encompasses monetarily steered success-oriented actions to a greater extent than other domains do; the economy exhibits forms of organization and types of interaction different from those found in family life; the actions constituting the economy make an important contribution to the reproduction of the material substrate and are linked monetarily to some of the actions constitutive of family life and public administration; and so on. I have further acknowledged the general propriety of model builders relying on generalizations in constructing predictive models of social affairs.

But not only is systems theory unnecessary to register these observations (after all, they are simply empirical generalizations), but there are many *other* generalizations about the economy that are not stylized as defining truths in action-systems theoretical (and other similar) depictions of economic affairs – for example, that business and consumer decisions are often made under intense pressure or off the cuff and are "irrational" and "emotional" (in Weber's sense); that many economic actions (e.g., buying) are something other than success-oriented actions; that the actions helping to constitute an economy count among their achievements not just maintenance of the material substrate, but also, inter alia, profit, self-aggrandizement, pleasure and satisfaction, promotion in corporate hierarchies, and, as Marx saw, the destruction of existing structures;[39] that vast systems of technological connections among settings link the actions constituting the economy; and that "organizational culture" (mostly commonalities but also orchestrations in intelligibility and setting) is a key feature of some successful economic firms. Whether generalizations of this sort add up to a "theory" of the economy (ontological, explanatory, or predictive) is a different question. What is clear is that Habermas's (and Parsons's) theories are just too preemptively simple for social ontological purposes. As I quoted Foucault in Chapter 1 on the theorization of power, an adequate ontological account of the modern economy can be achieved only by first investigating the intricacy and variety of the constellations of practices and *Zusammenhänge* of lives that constitute an economy, and then conducting an ascending analysis that achieves an overview of economic affairs by working through and climbing above this complexity.

A fourth task facing a more complete practice ontology is analysis of a phenomenon that has largely gone ignored in the current work, social change. Two dimensions of social change were described in the foregoing: (1) the expansion of intelligibility via novel actions and products, and (2) the opening up (via practice organizations) of acceptable or correct actions that have not yet been performed. These two dimensions, however significant, add up to considerably less than either a description or account of social transformation. For example, there is obviously more to social change (e.g., changes in government, evolution in race relations, and decaying infrastructures) than new and novel actions alone. Moreover, these dimensions feature individuals as the instigators of change and practices as the formative context in which they effect it. Is, however, the source of change rightly located exclusively in individuals? Might practices and nexuses thereof also be a site where change originates and maybe also is effected? And even if people are in some sense the sole origins and effectors of social transformation, might they be such in different ways (e.g., individually and collectively, intentionally and unintentionally), with practices constituting formative contexts differently in the various cases? Picturing the social field as a nexus of practices merely identifies the arena of transformation as masses of lives proceeding and hanging together within and across myriad practices. The implications of this conception for an adequate account of social change remain to be determined.

A fifth general task, one of a different nature than the preceding four, arises from the thesis, based on the practice theoretical depiction of the social field as a nexus of practices, that the sociality of social formations and phenomena is to be analyzed by reference to this nexus. This thesis implies that any proposed social ontological concept that captures something real about social life, and any alleged social phenomenon that actually exists, is either explicable in terms of the sociality established in nexused practices or captures a further detail of life and sociality in them. The first of these analyses applies to the fundamental ontological dimensions and types of social formation developed in competing social theories, insofar as they are valid determinations. Attempting to explicate thusly and hence evaluate the fundamental concepts of other theories is a subsidiary, though hardly insignificant task for a practice theoretical ontology.

As way of example, consider Parsons's, Habermas's, and early Luhmann's conception(s) of a symbolic medium of communication. According to these theorists, some such medium connects the actions that compose a given subsystem of society. If these conceptions capture something real about the social, media of communication will be explicable via the forms of sociality discussed in the previous two sections. I suggested earlier that these media might not be distinct from mind/actions and settings. This suggestion can now be more fully formulated as follows: a symbolic medium of communication is a general form of structuring of understandings, rules, teleoaffective orders, and layouts as well as connections among settings, which is such that the transmission of information or traces leads people to perform certain actions. What these three theorists capture in their conception(s) of a medium of communication are complex configurations of the specific forms of sociality that occur within and among practices. As argued earlier with reference to Habermas, however, these configurations are not such as to link actions into "systems" in any rigorous sense.

This practice theoretical rendering of a systems theoretical notion is made more perspicuous through the further explication, that what the idea of a symbolic medium of communication actually gestures at is the effectuation of influence. By "influence" I mean one person's actions intentionally or unintentionally determining how others act. One person's actions are able to determine another's only if information about or traces of them are transmitted to the other. This transmission occurs over perception, chains of action, and physical connections among settings. The propagation of influence can thus take many forms.

One experientially salient form occurs in face-to-face interaction. Face-to-face interactions are chains composed of actions that different people perform in one another's presence. What makes the actions into a chain is that they are responses to one another. In face-to-face interactions, as a result, actors influence one another by performing actions in one another's presence that constitute new situations for the others; and information that a particular action has been performed is transmitted via perception.

Another salient form of the propagation of influence is non-face-to-face interaction over spatial distance, for instance, conversation over the Internet. Here, actors influence one another by performing speech acts, copies of whose utilized words or symbols are produced via physical connections on other actors' monitors, thereby altering the latter's situations. Whenever, as in this case, information and traces are transmitted between settings that are not physically continuous, a physical connection serving as a medium of communication (e.g., electromagnetic waves or hard wiring) links the settings involved. Of course, the propagation of influence can also be effected by chains that continue for great lengths and that weave in and out of the same and different, spatially contiguous or discontinuous settings. In all instances, however, influence is propagated from one actor to another through perception, chains of action, and/or physical media of communication, as mediated by intelligibility and thus by orchestrations of and commonalities in intelligibility and settings. Action-systems theoretical media of communication are simply significant general forms taken by the conditions under which influence is propagated.

A sixth and final problem facing a social ontology based on the picture of the social field as a weave of practices is analysis of a phenomenon twice broached earlier: being one of us. That human beings are configured into diverse but overlapping sets of us and not-us has long been recognized in social thought as a prominent feature of social life. In Chapter 3, I defined a notion of being one of us via intelligibility. This notion was reformulated in Chapter 4 as the idea of being someone who participates in our practices, where "we" are the people participating in a particular set of dispersed practices interwoven into integrative ones. The present discussion permits further elaboration of this notion.

Someone who participates in a particular range of practices thereby coexists in a variety of ways with other participants. Accordingly, being one of us involves being party to the tissue of coexistences enmeshing us. It thus embraces, first, being caught up in the tissues opened in our dispersed practices. These consist centrally, though not at all exclusively, of commonalities in (1) actions, including speech acts; (2) promptings of and reactions to actions; and (3) comprehension of actions and what they express. Being one of us embraces, second, being caught up in the tissues of sociality opened within the linked integrative practices into which our dispersed practices are woven. The sociality enmeshing us in this dimension comprises a densely interwoven mat of the various types of coexistence discussed earlier in this chapter.

The limits this analysis places on a we approximate the boundaries of communities in Eugene Fink's characterization. Fink writes that in sharing a *Brauch* (custom), people are by and with one another in carrying out a common life in a shared world. This togetherness, a shared being-in-the-world, signals an elemental form of community, "the commonness of a community that is gathered and held in a world."[40] Each *Brauch* in this sense is a shared space of meaning and action, within which practitioners live with and understand one another in a single world. Although this formulation perpetuates the compositional emphasis on commonalities at the expense of orchestration and asymmetry, Fink's interpretation of a *Brauch* closely parallels the analysis of practice in this book. When linked, furthermore, *Bräuche* constitute a wider *Sittlichkeit* of interrelated being-in interconnected worlds (my construction) that signals a broader form of community: the establishment of a we composed of those interrelatedly gathered and held in interconnected worlds. Being a member of such a community, being party to an interrelated being-in interconnected worlds, closely resembles participation in shared dispersed practices interwoven into integrative ones. As Gurwitsch writes *avant la lettre* about communities in Fink's sense, "in advance, a human being is not solus ipse; insofar as he is communalized and historicalized, he always already belongs to other human beings."[41] Expressed in partly Fink-like terms, being one of us is being party to our conceptually infused and linguistically active interrelated being-in interconnected worlds.

This account of being one of us offers a contrast to that stream of theorizing which, from Simmel to most recently Margaret Gilbert, analyzes we-ness via shared psychological conditions focused on the we involved. On this line of thinking, being one of us is possessing (under the right conditions) a "we" thought, feeling, experience, belief, or intention. I have no dispute in principle with this tradition of analysis. The shared possession of such conditions *is* a constitutive feature of certain types of social formation (e.g., groups). Conflicts would arise only if representatives of this tradition were to claim (which they generally do not) that the sharing of "we conditions" is the sole type of we-ness configuring social life. Other forms of we-ness also clearly exist, for instance, the one intuitively associated with membership in a culture-sharing community engaged in diverse walks of life. The account outlined here sharpens this intuitive notion. By this juncture, moreover, it should be clear that this contextual sense of we-ness is just as significant in social life as – and in fact presupposed by – the intensional sense. As Wittgenstein writes about the certainty of background obviousnesses, " 'We are quite sure of it' does not mean just that every single person is certain of it, but that we belong to a community which is bound together by science and education" (*OC*, 298).

Postscript
Individual and Totality

This essay began by describing recent challenges to the two master concepts that, from its inception, have dominated thought about the nature of social life: individual and totality. I presented practice theory as one prominent stream of thought pressing these challenges and explained that its refusal to abandon general theory is one feature differentiating it from some contemporaneous movements critical of the traditional dualism. As we have since seen, two principal phenomena frame its attempt to fashion large-scale accounts of social life. The first is practices, the second understanding/intelligibility. Although practice theories tender divergent conceptions of practice, agreement reigns that practices are the central moment in social life because they are the site where understanding is carried and intelligibility articulated. Social life is an intricate nexus of practice, thus centrally a complicated weave of constellations of normativized understanding and intelligibility articulated through action.

If the picture of social life as a nexus of practice offers a genuine alternative to accounts organized around the notions of individual or totality, it should advance critique of the hegemonic claims traditionally raised for individuals and totalities in the theoretical study of social life. Analyses grounded on it should also underwrite more clairvoyant understandings of both individuals and whatever social wholes do exist. The current book can claim to have fulfilled these desiderata for individuals alone. As explained, practices constitute individuals by above all instituting their actions and mental as well as cognitive conditions; and they pull off this feat by housing and determining the contexts and understandings that are constitutive of the expressive relation between bodily activity and mind/action. The account developed in this book thus enriches understanding of individuality by explicating the social nature of its central dimension. In describing the social phenomena that the existence of individuals (i.e., mind/action) presupposes, it also furthers refutation of the foundational utility of the individual in general social ontology. This widely proclaimed but exaggerated utility was directly challenged in the concluding chapter, where the material with which individualist ontologies reconstruct social phenomena was shown to implicate a wider and omnipresent realm of practices and practice sociality.

My exposition has only gestured, however, at the implications of its "Wittgensteinian" account of practice for the possibility and character of social wholes. Proffered in this regard was a merely preemptive critique in Chapter 6 of one variant of a prominent contemporary notion of totality, that of a system of actions. The submitted criticisms, moreover, were preliminary and hand waving. A more adequate and thorough dissection of the ontological pretensions of totality awaits a more exhaustive account of social phenomena based on the conception of the social field as a nexus of practices. Only then can it be determined which – if any – sorts of whole characterize social life and whether any of the traditional notions of whole do them justice. Unlike individuals, who clearly exist, and whose character is tied to the nature of the fundamental constituent of the social field, totalities suffer a questionable and disputed existence, one that can be abjudicated and its character ascertained only with a more comprehensive account of this field.

Note, in addition, that this essay has generally avoided questions concerning the investigation of social life. Any account of the social has massive implications, however, for the nature of studying social phenomena. Depicting the social field as a nexus of integrative and dispersed practices, and these practices as manifolds of doings and sayings linked by understandings, rules, and teleoaffective structures, has significant implications for the cognitive achievements of social studies, the conceptual apparatuses and theories with which these can be realized, the exploration of social life and the gathering of information about it, as well as a variety of other issues. Epistemology, consequently, is a second general project commenced but unexecuted by the current document.

Notes

Chapter 1

1. This occurred even though there are no hard and fast implications between ontology and politics. Prior to the rise of socialism, for instance, the combination of liberalism and individualism first had to do battle against a coalition of totality ontologies and conservatism of the Edmund Burke and Adam H. Müller variety. For discussion, see Karl Mannheim, "Conservative Thought," in *From Karl Mannheim*, ed. Kurt H. Wolff, New York, Oxford University Press, 1971, pp. 132–222.

2. Derek Gregory, "Areal Differentiation and Post-Modern Human Geography," in *Horizons of Human Geography*, ed. Derek Gregory and Rex Walford, Totowa, N.J., Barnes and Noble, 1989, p. 69.

3. Zygmunt Bauman, *Intimations of Postmodernity*, London, Routledge, 1992, p. 189.

4. Michel Foucault, "Two Lectures," tr. Kate Soper, in Michel Foucault, *Power/Knowledge*, ed. Colin Gordon, New York, Pantheon, 1980, p. 99.

5. See Michel Foucault, "Nietzsche, Genealogy, History," in *The Foucault Reader*, ed. Paul Rabinow, New York, Pantheon, 1984, pp. 76–100, e.g., pp. 81, 89.

6. Anthony Giddens, *The Constitution of Society*, Berkeley, University of California Press, 1984, pp. 164–165.

7. Michael Mann, *The Sources of Social Power, Volume I: A History of Power from the Beginning to A.D. 1760*, Cambridge, Cambridge University Press, 1986, p. 1. Each network of power, furthermore, is organized nexuses of social interaction.

8. An argument to this effect, anticipating the communitarian assault, is found in Maurice Mandelbaum, "Social Facts," *British Journal of Sociology* 6 (1955), pp. 305–317.

9. See Jacques Lacan, *Écrits: A Selection*, tr. Alan Sheridan, London, Tavistock, 1977, and Julia Kristeva, *Revolution in Poetic Language*, tr. Margaret Waller, New York, Columbia University Press, 1984.

10. For elaboration, see Chantal Mouffe, "Feminism, Citizenship, and Radical Democratic Politics," in *Feminists Theorize the Political*, ed. Judith Butler and Joan Scott, London, Routledge, 1992, pp. 369–384.

11. A, for this writer, poignant illustration of this fact is found in Mike Leigh's film, *Four Days in July*. At the conclusion of this story about contemporary Northern Ireland, two mothers, a Catholic and a Protestant, relax in adjoining beds of a maternity ward after giving birth. At first they blissfully share the joys that are their newborn sons. The felicity is broken when the one asks the other for the name of her son, divulgence of which reveals the religious difference separating them. Further conversation ceases, and a sense of anomie envelops the scene.

12. This way of putting things is of course influenced by Richard Rorty, *Philosophy and the Mirror of Nature*, Princeton, Princeton University Press, 1979.

13. This theme runs through a number of the contributions to a recent book on postmodern social theory, *Postmodernism and Social Theory*, ed. Steven Seidman and David G. Wagner, Oxford, Blackwell, 1992. See, e.g., the otherwise careful and thoughtful treatment in Craig Calhoun, "Culture, History, and the Specificity of Social Theory," pp. 244–288.

14. The expression "practice theory" has gained currency in some disciplines, notably anthropology, where it is used primarily to designate the work of Bourdieu and Giddens. See, e.g., Sherry Ortner, "Theory in Anthropology since the Sixties," *Comparative Study of Society and History* 16 (1984), pp. 126–166, sect. 3; also Catherine Bell, *Ritual Theory, Ritual Practice*, Oxford, Oxford University Press, 1992, chap. 4.

15. These words are borrowed from Martin Heidegger, *Being and Time*, tr. John Macquarrie and Edward Robinson, Oxford, Blackwell, 1978, sec. 18 and 34.

16. Jean-François Lyotard, *The Postmodern Condition: A Report on Knowledge*, tr. Geoff Bennington and Brian Massumi, Minneapolis, University of Minnesota Press, 1979, p. xxiv.

17. Giddens, *The Constitution of Society*, p. xxii.

18. In addition to the works by Lyotard and Giddens already cited, see David Bloor, *Wittgenstein: A Social Theory of Knowledge*, New York, Columbia University Press, 1983, and Ernesto LaClau and Chantal Mouffe, *Hegemony and Socialist Strategy: Towards a Radical Democratic Politics*, London, Verso, 1985. For a general characterization of Wittgenstein's (post-Bloorian) impact on social studies of science, see Michael Lynch, "Extending Wittgenstein: The Pivotal Move from Epistemology to the Sociology of Science," in *Science as Practice and Culture*, ed. Andrew Pickering, Chicago, University of Chicago Press, 1992, pp. 215–265; and more recently, Michael Lynch, *Scientific Practice and Everyday Action*, Cambridge, Cambridge University Press, 1993, chap. 5. For a further constructive use of Wittgenstein in the sociology of knowledge, see Jeff Coulter, *Mind in Action*, Atlantic Highlands, N.J., Humanities Press International, 1989.

19. Talcott Parsons, *The Structure of Social Action*, New York, Free Press, 1968, p. 363; cf. p. 360.

20. This notion of order converges with that outlined by Bernhard Waldenfels in *Ordnung im Zwielicht*, Frankfort am Main, Suhrkamp, 1987, chap. B. According to Waldenfels, an order is a "regulated *Zusammenhang* between this and that, whose content [the nexus] is to be differentiated from its structure [the regulation]" (137). I write "converges with" instead of "is the same as" because I construe (social) order not, à la Waldenfels, as *Zusammenhänge* of lives, but instead as arrangements of lives therein implicated. More significant differences emerge in the details. For instance, Waldenfels maintains that all regulation proceeds by selection and exclusion, a claim too narrow for my tastes. Following Husserl and Schütz, moreover, he exhaustively locates the *Zusammenhänge* relevant to human existence in the experiential fields into which people act and speak. (These fields are articulated around themes determined by relevance, importance, and typicality.) This means that the "this" and "that" comprising an order's contents are experiential entities only, including actions and utterances.

21. Parsons, *The Structure of Social Action*, p. 91.

22. For a recent example, which also recognizes the role of norms, see Jon Elster, *The Cement of Society*, Cambridge, Cambridge University Press, 1989; cf. the introduction.

23. The significance of understanding/intelligibility in this regard is, how-

ever, acknowledged in strands of both ethnomethodology and symbolic interactionism. See the classical statements in Harold Garfinkel, *Studies in Ethnomethodology*, Englewood Cliffs, N.J., Prentice-Hall, 1967, and Harold Blumer, *Symbolic Interactionism*, Englewood Cliffs, N.J., Prentice-Hall, 1969.

24. Bauman, *Intimations of Postmodernity*, p. 196. Further page references are found in the text.

25. For discussion of this logic, see Laclau and Mouffe, *Hegemony and Socialist Strategy*, chap. 3.

26. My discussion will focus on texts of Wittgenstein from the period between (and including) the *Philosophical Investigations* and his death. I choose this focus in part because this is the period when he focused most intensely on mind/action and in part to avoid questions of development and continuity in his work. For an excellent discussion of the latter, see David G. Stern, *Wittgenstein on Mind and Language*, Oxford, Oxford University Press, 1994.

Chapter 2

1. An earlier, shorter, and somewhat different version of this chapter appeared as "Wittgenstein: Mind, Body, and Society," *Journal of the Theory of Social Behavior* 23, no. 3 (September 1993), pp. 285–314.

2. I have in mind here above all eliminative materialists. See, e.g., Richard Rorty, "Mind-Body Identity, Privacy, and the Categories," *Review of Metaphysics* 19, no. 1 (1965), pp. 24–54, and Paul Churchland, "Eliminative Materialism and the Propositional Attitudes," *Journal of Philosophy* 78 (1981), pp. 67–90.

3. Of course, not all behaviorisms are reductive in this sense. In the context of interpreting Wittgenstein, see, e.g., Willard F. Day, "On Certain Similarities between the *Philosophical Investigations* of Ludwig Wittgenstein and the Operationism of B. F. Skinner," *Journal of the Experimental Analysis of Behavior* 12, no. 3 (May 1969), pp. 489–506.

4. Thus, what the body "expresses" is neither independently existing entities such as "internal" psychological states nor ideal intellectual contents. These two prominent conceptions of what behavior "expresses" are best exemplified in Dilthey's early psychologistic and later post-Husserlian logical phases respectively. See Wilhelm Dilthey, *Gesammelte Schriften*, vol. 1 versus vol. 7, Leipzig, B. G. Teubner, 1927.

5. See, e.g., Allan Janik, "Wie hat Schopenhauer Wittgenstein beeinflusst?" *Schopenhauer Jahrbuch* 73 (1992), pp. 69–77; vis-à-vis Wittgenstein's earlier work, also P. M. S. Hacker, *Insight and Illusion*, 2d ed., Oxford, Clarendon Press, 1986, chap. 3.

6. On this, see David G. Stern, "Heraclitus' and Wittgenstein's River Images: Stepping Twice into the Same River," *Monist* 74, no. 4 (October 1991), pp. 581–604.

7. This latter formulation is Heidegger's "formal" definition of a phenomenon. See Martin Heidegger, *Being and Time*, tr. John Macquarrie and Edward Robinson, Oxford, Blackwell, 1978, pp. 51–55. The notion of an arena in which unvarnished reality presents itself unites – and is partly definitive of – twentieth-century phenomenology. The connections of Husserl's, Heidegger's, and Merleau-Ponty's conception of a phenomenal realm with Wittgenstein's use of the terms *Erscheinungen* and *Phänomene* cannot be further explored here. Unfortunately, this parallel is not examined in the one book that compares Wittgenstein to these thinkers in detail, Nicholas F. Gier,

Wittgenstein and Phenomenology: A Comparative Study of the Later Wittgenstein, Husserl, Heidegger, and Merleau-Ponty, Albany, State University of New York Press, 1981.

8. For further discussion of this point, see my article "Wittgenstein + Heidegger on the Stream of Life," *Inquiry* 36, no. 3 (September 1993), pp. 307–328.

9. In the article mentioned in the previous footnote, I translated *Zustand* as "state." Peter Winch's verbal objections to the use of this English word in connection with Wittgenstein led me to reconsider its propriety.

10. See Norman Malcolm, "Wittgenstein on the Nature of Mind," in his *Thought and Knowledge,* Ithaca, N.Y., Cornell University Press, 1977, pp. 133–158, and David Sachs, "Wittgenstein on Emotion," *Acta Philosophica Fennica* 28 (1976), pp. 250–285.

11. Rom Harré, *Physical Being,* Oxford, Blackwell, 1992, p. 142.

12. Peter Strawson, *Individuals,* London, Methuen, 1959, chap. 3. The class of condition-of-life "predicates" is closer to the class of Shoemaker's P*-predicates than to that of Strawson's P*-predicates. See Sydney Shoemaker, "Self-Reference and Self-Awareness," *Journal of Philosophy* 65, No. 19 (October 1968), pp. 555–567.

13. This position parallels Dennett's, who claims that personhood is defined by the possession of a series of attributes, exemplification of some of which, e.g., self-consciousness and the ability to communicate verbally, presuppose the possession of others, e.g., rationality and intentionality, that is, conditions of life. See Daniel Dennett, "Conditions of Personhood," in *The Identities of Persons,* ed. Amélie Oksenberg Rorty, Berkeley, University of California Press, 1976, pp. 175–196.

14. Judith Butler, *Gender Trouble: Feminism and the Subversion of Identity,* New York, Routledge, 1990, pp. 16, 24.

15. See, e.g., Donald Davidson, "Mental Events," in *Experience and Theory,* ed. Lawrence Foster and J. W. Swanson, London, Duckworth, 1970, pp. 79–102.

16. For a parallel but different classification, see Malcolm Budd, *Wittgenstein's Philosophy of Psychology,* New York, Routledge, 1989, chap. 1.

17. Although he is not consistent, in a large number of cases Wittgenstein employs *seelisch* to refer to mental conditions (categories one and two) and *geistig* to designate cognitive ones (category three).

18. The loci classici for this analysis are Arthur Danto, "Basic Actions," *American Philosophical Quarterly* 2, no. 2 (April 1965), pp. 141–48, and Alvin Goldman, *A Theory of Human Action,* Princeton, Princeton University Press, 1970, chap. 2 and 3.

19. See, e.g., Donald Davidson, "Belief and the Basis of Meaning," *Synthese* 27 (1974), pp. 309–323.

20. Wittgenstein never explicitly characterizes sensations and images thusly. But if one acknowledges the existence of phenomena of pain and imagery – as opposed to the conditions of being in pain and of imagining – and equates phenomena with appearances, the characterization follows when one asks to whom such phenomena are appearances.

21. For more discussion of this point, see my "Wittgenstein + Heidegger on the Stream of Life."

22. Medard Boss, *Existential Foundations of Medicine and Psychology,* tr. Stephen Conway and Anne Cleaves, New York, Jason Aronson, 1979, pp. 102–103.

23. Interestingly, this same notion of ways of being, accompanied not only by the idea that these ways are designated by common locutions of the sorts just mentioned but also by the claim that the phenomena that express them are the realities of life behind which no substance, ground, or reality lies is found in Helmut Plessner, *Laughing and Crying: A Study of the Limits of Human Behavior*, tr. James Spencer Churchill and Marjorie Grene, Evanston, Ill., Northwestern University Press, 1970, introduction.

24. This distinction, though not the particular rendering of it, is due to Plessner, ibid.

25. See Heidegger, *Being and Time*, sec. 18.

26. Butler, *Gender Trouble*, chap. 1, sec. 5, and chap. 3, sec. 4.

27. Ibid., pp. 135–136.

28. So defined, speech acts are only one of the kinds of speech act identified by Searle. Others include utterance, illocutionary, and perlocutionary acts. See John Searle, *Speech Acts*, Cambridge, Cambridge University Press, 1969, chap. 2.

29. For discussion, see Eike von Savigny, "Avowals in the *Philosophical Investigations:* Expression, Reliability, Description," *Nous* 24 (1990), pp. 507–527.

30. I realize that much has been written pro and con about Wittgenstein's asseverations on this topic. It would take up too much space, however, to engage with the literature here. I believe that Haig Khatchadourian still offers a clear and persuasive defense [*sic*] of Wittgenstein's point in "Common Names and Family 'Resemblances,'" *Philosophy and Phenomenological Research* 18 (1958), pp. 341–358.

31. For discussion, see Kjell S. Johannessen, "Rule Following, Intransitive Understanding, and Tacit Knowledge," in *Essays in Pragmatic Philosophy*, ed. Helge Hoibraaten, Oxford, Oxford University Press, 1990, pp. 101–127.

32. For additional discussion, see Hubert Dreyfus and Stuart Dreyfus, *Mind over Machine*, New York, Free Press, 1986.

33. In the context of interpreting Wittgenstein, see Peter Winch, *The Idea of a Social Science and Its Relation to Philosophy*, London, Routledge and Kegan Paul, 1958, p. 58.

34. Again in the context of interpreting Wittgenstein, see Norman Malcolm, "Wittgenstein on Language and Rules," *Philosophy* 64 (1989), p. 22. I believe that Malcolm comes closest to capturing what Wittgenstein means in those places where he implies that language use, or at least parts of it, is rule governed: To follow a rule is to use language, as it is normally used, in those domains of usage where significant regularity exists (e.g., *LW* I, 968–969). On this interpretation, however, to say that language is rule governed is simply to say that it is regular. So the comments in the text concerning unformulability and the consequent superfluity of rule talk apply. This means, I think, that Wittgenstein offers powerful arguments against a form of talk to which he himself inclines.

35. Maurice Merleau-Ponty, *The Phenomenology of Perception*, rev. trans., tr. Colin Smith, London, Routledge, 1989, p. 180.

36. A second interpretation of the preceding assertion (there are still others having to do with sensations and sense impressions) is provided by the criterionless self-ascription of life conditions. Wittgenstein insists that the ability to avow conditions of life must not be explained by invoking (inner) acts of recognition, exploration, and identification. Self-ascriptions are not *products* of prior acts, but *beginnings*, initiating moves in language-games with others. As Wittgenstein sometimes writes, they are signals (to others). Self-

ascription is thus both spontaneous, in not being based on a prior act of identification, and an originary act that initiates something new, namely, the ensuing language-game.

37. See Stephen Mulhall's discussion of humanity *qua* personal physiognomy in *On Being in the World: Wittgenstein and Heidegger on Seeing Aspects*, New York, Routledge, 1990, pp. 86–87.

38. Ludwig Klages, *Vom Wesen des Bewusstseins*, Leipzig, Johann Ambrosius Barth, 1933, p. 26.

Chapter 3

1. Of the above mentioned contemporaries, Wittgenstein was most familiar with the work of Freud and Spengler. For discussion of Wittgenstein's connection with the wave of conservative malaise typified by Spengler, see J. Nyiri, "Wittgenstein's Later Work in Relation to Conservatism," and Georg Henrik von Wright, "Wittgenstein in Relation to His Times," both in *Wittgenstein and His Times*, ed. Brian McGuinness, Oxford, Blackwell, 1982, pp. 44–68, and 108–120; also Rudolf Haller, *Questions on Wittgenstein*, Lincoln, University of Nebraska Press, 1988, chaps. 5 and 6, and David Bloor, *Wittgenstein: A Social Theory of Knowledge*, New York, Columbia University Press, 1983, chap. 8.

2. For development and defense of this thesis, see G. E. M. Anscombe, *Intention*, Ithaca, N.Y., Cornell University Press, 1957.

3. Wittgenstein is not the first thinker to highlight the significance of reactions. Similar emphases with social ontological import are found in John Dewey's notion of habit and Heidegger's notion of dealings (*Umgänge*) with equipment. See John Dewey, *Human Nature and Conduct: An Introduction to Social Psychology*, London, George Allen and Unwin, 1922, part one, and Martin Heidegger, *Being and Time*, tr. John Macquarrie and Edward Robinson, Oxford, Blackwell, 1978, sec. 13, 15–16.

4. See, e.g., Friedrich Nietzsche, *The Twilight of the Idols*, in *The Portable Nietzsche*, tr. Walter Kaufmann, New York, Penguin, 1982, "The Four Great Errors," sec. 3 and 7; and Friedrich Nietzsche, *Beyond Good and Evil*, tr. Walter Kaufmann, New York, Random House, 1966, p. 19.

5. As is well known, Wittgenstein repeatedly uses the word *Abrichtung* (training, with strong connotations of drill) to characterize a child's education in speaking and acting. In fact, of course, much of this education does not consist in adults explicitly instructing children. A child acquires a tremendous amount simply by existing amidst adults and their practices. I will speak, consequently, of training and learning instead of training alone.

6. This is demonstrated by the language learning abilities of autists. See, most recently, the case of the autist Birger Sellin, who relates that at age five he could and did read books in German, even though before reaching the age of two he had ceased reacting normally and had become completely disengaged from any sort of education children normally receive. Cf. Birger Sellin, *ich will kein inmich mehr sein, botschaften aus einem autistischen kerker*, ed. Michael Klonovsky, Cologne, Kiepenheuer & Witsch, 1993.

7. I have discussed Wittgenstein's views on the interpretive human sciences in "Elements of a Wittgensteinian Philosophy of the Human Sciences," *Synthese* 87 (1991), pp. 311–329. For discussion of his views on aesthetics, see Kjell S. Johannessen, "Language, Art, and Aesthetic Practice," in *Wittgenstein – Aesthetics and Transcendental Philosophy*, ed. Kjell S. Johannessen and Tore Nordenstam, Vienna, Verlag Hölder-Pichler-Tempsky, 1981, pp. 108–126,

and Benjamin R. Tilghman, *But Is It Art? The Value of Art and the Temptation of Theory*, Oxford, Blackwell, 1984, chap. 3 and 6.

8. Wittgenstein is reported as asserting this explicitly about dreams, play, and punishment in *Lectures and Conversations on Aesthetics, Psychology, and Religious Belief*, ed. Cyril Barrett, Oxford, Blackwell, 1966, pp. 47, 49, 50.

9. I write "psychoneurophysiological" to indicate that I do not want to prejudge or prescribe the form taken by causal explanatory investigation of conditions and expressions of life. The causality responsible for expressions of life is presumably lodged in the body's neurophysiologicalhormonal systems. Nothing in the Wittgensteinian account developed in this study implies, however, that theories about this causality must therefore be limited to overtly physicalistic categories. Hence the prefix "psycho." Of course, we also should not rule out the possibility of discovering a bodily system in addition to or other than neurophysiologicalhormonal ones that houses the relevant causality. I just don't think that Wittgenstein's account of mind/action has many implications for the causal explanatory study of the expressions of mind/action (apart from suggesting that common mind/action locutions will not be of use in this study unless they are respecified as technical terms). This claim converges with the identical contention about "Wittgensteinian philosophies" of mind found in Richard Rorty, "Wittgensteinian Philosophy and Empirical Psychology," *Philosophical Studies* 31 (1977), pp. 151–172. This article does not discuss Wittgenstein directly.

10. Thus I do not follow such interpreters of Wittgenstein as Norman Malcolm and Paul Feyerabend who argue that, for Wittgenstein, any human scientific enterprise that appropriates common locutions for mentality and activity, but uses them in ways differing from how they are commonly used (e.g., establishes new criteria for them), has thereby automatically "changed the subject" and is no longer using the same concept or talking about the same aspect of human life. (See Norman Malcolm, *Dreaming*, London, Routledge and Kegan Paul, 1962, and Paul Feyerabend, "Wittgenstein's 'Philosophical Investigations,'" *Philosophical Review* 64 [1955], pp. 449–83.) The understandings that are embodied in techniques can be influenced by theories in the human sciences, just as such theories constantly draw upon common understandings. Common understandings, moreover, are not monolithic, but instead gradually modulate across cultures (e.g., American versus Finnish versus Italian) and historical eras (e.g., fourth-century Florence versus Renaissance Tuscany versus twenty-fifth-century Umbria). No formula, consequently, can determine exactly when science is expressing new concepts in old words. A discussion between scientists and nonscientists can usually ascertain, however, whether this has taken place.

11. See, e.g., Max Black, "*Lebensform* and *Sprachspiel* in Wittgenstein's Later Work," in *Wittgenstein and His Impact on Contemporary Thought: Proceedings of the Second International Wittgenstein Symposium*, ed. E. Leinfellner et al., Vienna, Verlag Hölder-Pichler-Tempsky, 1978, pp. 325–331.

12. An example of the former is Henry Le Roy Finch, *Wittgenstein – The Later Philosophy: An Exposition of the "Philosophical Investigations*," Atlantic Highlands, N.J., Humanities Press, 1977, chap. 7; and of the latter, J. F. M. Hunter, "Forms of Life in Wittgenstein's Philosophical Investigations," *American Philosophical Quarterly* 5 (1968), pp. 233–243.

13. The consanguinity of the social and the biological in a form of life is nicely registered in Garver's thesis, that "the" human form of life is a phenomenon of natural history comprising the "common behavior of mankind" (*PI*, 206). Since this behavior includes activities as diverse as commanding,

questioning, recounting, eating, playing, and walking (*PI*, 25), the human form of life is clearly social-biological and not purely biological. Unfortunately, Garver denies that a form of life can also be something sociocultural and argues that what Wittgenstein exclusively means by the term is the human form of life. See Newton Garver, "Form of Life in Wittgenstein's Later Works," *Dialectica* 44, nos. 1–2 (1990), pp. 175–201.

14. A fact Hans-Georg Gadamer emphasizes in his hermeneutics; cf. *Truth and Method*, 2d, rev. ed., tr. Joel Weinsheimer and Donald G. Marshall, New York, Crossroad, 1989, part 3, chap. 2, sec. c.

15. This characterization of intransitive understanding derives from Kjell S. Johannessen, "Art, Philosophy and Intransitive Understanding," in *Wittgenstein – Eine Neubewertung*, ed. Rudolf Haller and Johannes Brandl, Vienna, Verlag Hölder-Pichler-Tempsky, 1990, pp. 323–333.

16. For discussion of parallels in Wittgenstein's views on understanding humans and moral (and aesthetic) judgment, see Benjamin R. Tilghman, *Wittgenstein, Ethics, and Aesthetics: The View from Eternity*, London, Macmillan, 1991.

17. Since the present point is that the expressive relation depends on conceptual understanding, I will not specify which (whose) understandings are the relevant ones. Chapter 4 will explain that understandings of life conditions are carried in practices and that the conceptual understandings against which a given behavior-in-circumstances expresses particular life conditions are mostly those carried by and woven into the nexus of practices in which the actor participates.

18. For a discussion of the further-reaching idea that certain meanings of certain bodily behaviors are rooted in biological evolutionary heritage, see Maxine Sheets-Johnstone's conception of corporeal archetypes: meanings built through primal heritage into such behaviors as staring, averting the eyes, turning one's back, spreading one's legs, and increasing one's height. Maxine Sheets-Johnstone, *The Roots of Power: Animate Form and Gendered Bodies*, Chicago, Open Court, 1994, chaps. 2 and 3.

19. The dependency of a person's condition(s) on conceptual understanding converges with Tyler Burge's well-known argument for the dependency of mental states on language use (or linguistic convention; Tyler Burge, "Individualism and the Mental," *Midwest Studies in Philosophy IV: Studies in Metaphysics*, ed. Peter A. French, Theodore E. Uehling, and Howard K. Wettstein, Minneapolis, University of Minnesota Press, 1979, pp. 73–121). Burge maintains that a "subject's belief or intention contents can be conceived to vary simply by varying conventions in the community about him" (p. 109). Assuming that different contents equal different states (pp. 75, 77), it follows that a person's mental states depends on social practices of language use. This position parallels the thesis that a person's conditions depend on conceptual understandings, since these understandings are prominently manifested in the use of words for conditions. My position amplifies Burge's in this regard by additionally resting a person's conditions on the accompanying patterns of nonverbal actions in which understanding is also expressed.

Key elements of the position developed here also appear in Noel Fleming's article, "Seeing the Soul," *Philosophy* 53 (1978), pp. 33–51. On Fleming's interpretation of Wittgenstein, which condition of life is "displayed" by a given piece of behavior depends on the behavior's circumstances along with the ways people generally react to one another's behavior. Because the reactions involved are actions, Fleming avoids reductive behaviorism. He does not, however, analyze the notion of "display," i.e., expression. This oversight, I believe, underpins a number of overly behavioristic specifications of life conditions, e.g., "In fact, the joy in someone's face is more fundamental for the concept of joy, or what joy is, than anything that goes on privately in anyone; because that – looking and acting like that, in circumstances like these – is what we call being joyful" (p. 36).

For further development of the idea that which psychological condition(s) activity expresses depends on social patterns of behavior, see Eike von Savigny, *Der Mensch als Mitmensch. Studien zu Wittgensteins Philosophische Untersuchungen*, Frankfurt am Main, Suhrkamp, 1995, chaps. 6–9, 11–12.

20. For a discussion of the indeterminacy and nonpregiveness of mind/action that parallels the present treatment, see Paul Johnston, *Wittgenstein: Rethinking the Inner*, London, Routledge, 1993, chap. 4.

21. This theme joins a large number of writers influenced by Wittgenstein (it also has of course a wider following and longer history), e.g., Peter Winch, *The Idea of a Social Science and Its Relation to Philosophy*, London, Routledge and Kegan Paul, 1958, chap. 3; A. Melden, *Free Action*, London, Routledge and Kegan Paul, 1967; and Georg Henrik von Wright, *Explanation and Understanding*, Ithaca, N.Y., Cornell University Press, 1971. A parallel but different version of this theme is found in A. Louch, *Explanation and Human Action*, Berkeley, University of California Press, 1966.

22. I have argued that responded-to phenomena are causes of action in "Social Causality," *Inquiry* 31, no. 2 (1988), pp. 151–70. Wittgenstein, too, sometimes writes this way, e.g., *RPP II*, 168.

23. For a cavalcade of recent contrasting analyses of seeing-as, see Malcolm Budd, *Wittgenstein's Philosophy of Psychology*, New York, Routledge, 1989, chap. 4; Stephen Mulhall, *On Being in the World: Wittgenstein and Heidegger on Seeing Aspects*, New York, Routledge, 1990, esp. chap. 1; and Johnston, *Wittgenstein: Rethinking the Inner*, chap. 2.

24. Jerry Fodor and Charles Chihara, "Operationalism and Ordinary Language," in *Wittgenstein: The Philosophical Investigations*, ed. George Pitcher, London, Macmillan, 1968, pp. 384–419. References to this article appear in the text.

25. Expressive bodies must not be wholly or even partly organic. Cyborgs can have and talk about their conditions of life, as well as ascertain them in others, provided that their behavior exhibits sufficient multiplicity and is sensitive to context. This is also imaginable vis-à-vis robots, although at the present level of technology their observable activities lack the features necessary to support mind/action.

26. Michel Foucault, "The Subject and Power," in Hubert L. Dreyfus and Paul Rabinow, *Michel Foucault: Beyond Structuralism and Hermeneutics*, 2d ed., Chicago, University of Chicago Press, 1983, p. 212.

27. See Michel Foucault, "The History of Sexuality," in *Power/Knowledge*, ed. Colin Gordon, tr. Leo Marshall, New York, Pantheon, 1980, p. 185; also Michel Foucault, *The History of Sexuality, Volume I*, tr. Robert Hurley, New York, Vintage, 1980, pp. 152–154.

28. This formulation stems from Judith Butler, *Gender Trouble: Feminism and the Subversion of Identity*, New York, Routledge, 1990, p. 18, and is compatible with the new ideas articulated in her more recent book, *Bodies That Matter: On the Discursive Limits of Sex*, London, Routledge, 1993. Although she does not focus on bodily enactment in that work, gender is still viewed as enmeshing the four phenomena mentioned in the formulation.

29. Butler, *Gender Trouble*, p. 140.

Chapter 4

1. For discussion, see Georg Henrik von Wright, *Explanation and Understanding*, Ithaca, N.Y., Cornell University Press, 1971, chap. 4, and my article, "Social Causality," *Inquiry* 31, no. 2 (1988), pp. 151–170. Bifurcating and intersecting causal chains are the central mechanism through which unintended consequences of action are brought about (even though chains of action, like consequences thereof, can and are sometimes intended by the people initiating or extending them).

2. See Jürgen Habermas, *The Theory of Communicative Action, Volume 2: Lifeworld and System: A Critique of Functionalist Reason*, tr. Thomas McCarthy, Boston, Beacon Press, 1987, p. 117.

3. See ibid., chap. 6, sec. 1B.

4. This holds only when there actually is a dispersed practice of X-ing. When there is no such practice, then for reasons that exactly parallel those presently outlined in the text, an act of X-ing presupposes the integrative practice that establishes the activity of X-ing. This subtlety will become clearer in the succeeding section.

5. I remind the reader that I analyze human actuality, not possibility. My claim is that *as things happen to stand*, the conceptual understanding of X-ing is carried in the practice of X-ing. I do not maintain that it is logically impossible for it to be otherwise, that, for instance, it is logically impossible that there should be created a solitary creature on a remote asteroid who understands X-ing. This means that arguments like those of Margaret Gilbert, which maintain against such Wittgensteinians as Peter Winch that understanding a concept does not necessitate a social context, pass my position by. (See Margaret Gilbert, *On Social Facts*, Princeton, Princeton University Press, 1989, chap. 3.) All the more so since Gilbert's considerations in this context are directed against the specific argument, that understanding a concept entails a social context because it entails the possibility of others being able to tell on the basis of someone's behavior what concepts he understands. My position is simply that people's understandings of action concepts as a matter of fact derive from practices (and representations thereof).

6. Remember that this applies where "X" is a common locution for activity. I am not denying that investigators might give words that designate actions technical meanings or invent technical vocabulary to describe what people do. Also remember that this conclusion, as formulated, applies only when the dispersed practice of X-ing actually exists.

I should also use this footnote to point out that, although for the sake of expository clarity I speak in the text of *the* dispersed practice of X-ing, I do not mean to exclude the existence of multiple such practices and thus multiple understandings of X-ing. Strictly speaking, consequently, my thesis is that X-ing presupposes *a* dispersed practice of X-ing. The following section will explain which of a plurality of X-ing practices – whenever such a plurality exists – is the one presupposed by a given X-ing.

7. Malcolm Budd, *Wittgenstein's Philosophy of Psychology*, New York, Routledge, 1989, pp. 39–42. Budd compiles other passages with parallel formulations. See also G. P. Baker and P. M. S. Hacker, *Wittgenstein: Rules, Grammar, and Necessity*, Oxford, Blackwell, 1986, p. 164.

8. For discussion, see Eike von Savigny, "Self-Conscious Individual versus Social Soul: The Rationale of Wittgenstein's Discussion of Rule Following," *Philosophy and Phenomenological Research* 51, no. 1 (March 1991), pp. 67–84.

9. I realize, of course, that emphatic assertion does not make it so. For arguments with which I am largely in agreement, see Norman Malcolm, "Wittgenstein on Language and Rules," *Philosophy* 64 (1989), pp. 5–28. Malcolm's remarks do not concern technique, but what it is to follow a rule, and they specifically oppose the "individualist" reading of Wittgenstein's views thereof propounded in Baker and Hacker, *Wittgenstein: Rules, Grammar, and Necessity.*

10. See also "Notes for the Philosophical Lecture," in Ludwig Wittgenstein, *Philosophical Occasions,* ed. James Klagge and Alfred Nordman, tr. David G. Stern, Indianapolis, Ind., Hackett, 1993, pp. 445–457.

11. Christopher Peacocke, "Reply: Rule-Following: The Nature of Wittgenstein's Arguments," in *Wittgenstein: To Follow a Rule,* ed. S. H. Holtzman and C. M. Leich, London, Routledge and Kegan Paul, 1981, pp. 72–95.

12. These latter sorts of action indicate what the use of the word accomplishes. Cf. *PI,* 156, where Wittgenstein writes that the language-game with a word reflects the role the word plays in our life.

13. Michael Oakeshott, *On Human Conduct,* Oxford, Clarendon Press, 1975, p. 55. Page references are henceforth in the text.

14. The distinction between (1) and (2) (which Oakeshott characterizes as one between self-disclosure and self-enactment) is more or less Alfred Schütz's distinction between in-order-to and because motives; in more colloquial English, the unstable difference between purpose/intention and reason/motive. See Alfred Schütz, *The Phenomenology of the Social World,* tr. George Walsh and Frederick Lehnert, Evanston, Ill., Northwestern University Press, 1967, sec. 17 and 18.

15. Notice (cf. note 4), therefore, that individual acts of Q-ing and R-ing do not presuppose the dispersed practices of Q-ing and R-ing, but instead the integrative practice that establishes Q-ing and R-ing.

16. That temporal pace and rhythm help compose a practice and are part of what is grasped in understanding it is brilliantly demonstrated by Pierre Bourdieu in his analysis of gift giving in Kabylia. See *Outline of a Theory of Practice,* tr. Richard Nice, Cambridge, Cambridge University Press, 1976, chap. 1, sec. 1.

17. Hubert Dreyfus, *Being-in-the-World: A Commentary on Heidegger's Being and Time, Division I,* Cambridge, Mass., MIT Press, 1991, chap. 4, sec. 2.

18. The centrality of normativity to practices is stressed in Dreyfus's interpretation of Heidegger's notion of *das Man.* See ibid., chap. 8, sec. II; and Martin Heidegger, *Being and Time,* tr. John Macquarrie and Edward Robinson, Oxford, Blackwell, 1978, sec. 27.

19. Charles Taylor, "Interpretation and the Sciences of Man," in his *Philosophy and the Human Sciences: Philosophical Papers 2,* Cambridge, Cambridge University Press, 1985, p. 36.

20. Ibid.

21. Stephen Turner, *The Social Theory of Practices: Tradition, Tacit Knowledge and Presuppositions,* Cambridge, Polity Press, 1994. References to this book are herewith found in the text.

22. Turner obliquely acknowledges the possibility of noncausal, nonsubstantial accounts of preferences, purposes, values, and the like by distinguishing between instrumental and realist (causal) construals of them. He brusquely rejects the instrumental position, however, on the grounds that it neglects "the need to connect the stuff of thought to the world of cause and substance" (p. 37). This is precisely the connection that Wittgenstein, though not instrumentalists generally, rejects.

NOTES 223

23. This is not to say that all people who share understandings act mutually intelligibly. It does mean, however, that the correct description of a situation where, after years of mutually intelligible X-ings, someone X-s unintelligibly to the others involved, is that their understandings of X-ing have begun to diverge after having been the same. On a causal construal of understanding, the proper description would be that they have finally discovered that their understandings had actually differed all that time. Whatever counterintuitiveness vexes the Wittgensteinian description arises from its noncausal rendering of understanding.

24. As W. V. Quine urged, meanings are not free-standing, distinct entities. They exist, as will quickly emerge, only in human understanding. See Quine's evocation of the "museum myth" of meaning in "Ontological Relativity," in his *Ontological Relativity and Other Essays*, New York, Columbia University Press, 1969, p. 27.

25. Heidegger, *Being and Time*, sec. 34.

26. Cf. Martin Heidegger, "Letter on Humanism," in *Basic Writings*, ed. David Krell, New York, Harper and Row, 1977, p. 193.

27. The reader may recognize the influence upon the last two paragraphs of Heidegger's analysis of equipment and the equipmental whole (*Zeugganz*). See Heidegger, *Being and Time*, sec. 15–16. See also the superb discussion in Dreyfus, *Being-in-the-World: A Commentary on Being and Time, Division I*, chap. 5, sec. 3.

28. See Heidegger, *Being and Time*, sec. 22.

29. For a convergent and similarly Heideggarian discussion of equipment, place, and action, see Jakob Meløe, "The Agent and His World," in *Praxeology: An Anthology*, ed. Gunnar Skirbekk, Oslo, Universitetsforlaget, 1983, pp. 13–29.

30. For a more extensive analysis of the space of places, its relation to objective space, and its relevance for social ontology and explanation, see my "Spatial Ontology and Explanation," *Annals of the Association of American Geographers* 81, no. 4 (1991), pp. 650–670. See also the parallel views in Anthony Giddens, "Time, Space, and Regionalization," in *Social Relations and Spatial Structures*, ed. Derek Gregory and John Urry, London, Macmillan, 1985, pp. 265–295. For one of the few social theoretical treatises that treats social space as not merely an objective space, but also a space of places, see Edward Soja, *Postmodern Geographies*, London, Verso, 1989, chap. 6.

31. Laclau and Mouffe reject "the distinction between discursive and nondiscursive practices" (Ernesto Laclau and Chantal Mouffe, *Hegemony and Socialist Strategy: Towards A Radical Democratic Politics*, London, Verso, 1985, p. 107). I do not believe, however, that they thereby deny the difference between linguistic and nonlinguistic behavior, only the existence of a sphere of behavior beyond discursiveness – where "discursiveness" means articulated intelligibility, not language. What they reject, consequently, is action that lies "outside" the sites where intelligibility is articulated (discourses). This thesis parallels my claim that actions presuppose practices.

32. Ibid., p. 113.

33. G. E. M. Anscombe, *Intention*, Ithaca, N.Y., Cornell University Press, 1957, sec. 23–26.

34. Ibid., p. 80.

35. Dreyfus, *Being-in-the-World: A Commentary on Heidegger's Being and Time, Division I*.

36. See also Hans Joas's claim that motives, ends, intentions, and the like are the products of reflection upon a prereflective *Ausgerichtetheit* of bodily

activity, a claim that rests on the same two assumptions as Dreyfus's position. Cf. Hans Joas, *Die Kreativität des Handelns*, Frankfurt am Main, Suhrkamp, 1992, p. 237.

37. Dreyfus, *Being-in-the-World: A Commentary on Heidegger's Being and Time, Division I*, p. 85.

38. Oakeshott, *On Human Conduct*, p. 89. Further page references are in the text.

39. This sensibility is most famously expressed in a number of articles, especially "The Tower of Babel" and "Rational Conduct" in his *Rationalism in Politics and Other Essays*, London, Methuen, 1962, pp. 59–79 and 80–110.

40. For greater explication and contextualization of the following interpretation of Heidegger's notion of signifying, see my "Wittgenstein + Heidegger on the Stream of Life," *Inquiry* 36, no. 3 (September 1993), pp. 307–328. I explain there that Heidegger does not present his views on understanding and *Befindlichkeit* as a contribution to an analysis of action, and that my interpretation, as suggested by the translation of *Befindlichkeit* as "attunement," abstracts from many of his central concerns in *Being and Time*. I also try in that article to explain in detail why Wittgenstein's account of life conditions and Heidegger's description of signifying are not merely compatible with one another but, in an important sense, coextensive. For a propitious account of understanding and *Befindlichkeit* that construes their relation to action differently, but not too differently from how it is sketched here, see Dreyfus, *Being-in-the-World: A Commentary on Heidegger's Being and Time, Division I*, chap. 10 and 11.

41. As described in the previous chapter, language's failure to convey meanings adequately is equally flagrant in the aesthetic realm, where the understood phenomenal presence of musical pieces, works of art, dance numbers, and architectural works is richer and more nuanced than what of it can be put into words. I will not, however, consider this additional frontier where the limits of language fall short of the limits of intelligibility. For discussion, see Kjell S. Johannessen, "Art, Philosophy and Intransitive Understanding," in *Wittgenstein – Eine Neubewertung*, ed. Rudolf Haller and Johannes Brandl, Vienna, Verlag Hölder-Pichler-Tempsky, 1990, pp. 323–333.

42. This point resembles Theodor Adorno's claim that the particularity of a particular cannot be grasped in concepts. It might be said in the present context that if anything can mark particularity, reactions, not words, do so. See Theodor Adorno, *Negative Dialectics*, tr. E. B. Ashton, New York, Continuum, 1973, part 2.

43. Dreyfus, *Being-in-the-World: A Commentary on Heidegger's Being and Time, Division I*, p. 215.

44. See Maja-Lisa Perby, "Computerization and Skill in Local Weather Forecasting," in *Knowledge, Skill and Artificial Intelligence*, ed. Bo Göranzon and Ingela Josefson, Berlin, Springer-Verlag, 1988, pp. 39–52.

Chapter 5

1. I should make clear from the start that, because my present interest concerns analyses of practice, my encounter with Bourdieu and Giddens grapples primarily with their abstract, theoretical accounts of practice and does not attempt to analyze either the remainder of their theoretical ideas nor the relation of these accounts to their more empirical investigations of various facets and domains of social life. For a discussion of the latter issue

vis-à-vis Bourdieu, see Derek Robbins, *The Work of Pierre Bourdieu*, Milton Keynes, Open University Press, 1991, chap. 8 and 9. I further stress that my exposition will highlight certain dimensions of their accounts, above all practical understanding, while deemphasizing other dimensions, for example, power. As we shall see, their accounts of practical understanding underlie much of the remainder of their theories.

2. At one point, for instance, he designates the linking of phrases as the domain of politics and adds that politics is "the state of language" (*The Differend: Phrases in Dispute*, tr. Georges van den Abbeele, Minneapolis, University of Minnesota Press, 1988, sec. 190). Numerous others sections similarly equate phrases implicatively with linguistic utterances/statements.

3. Jean-François Lyotard, *The Postmodern Condition: A Report on Knowledge*, tr. Geoff Bennington and Brian Massumi, Minneapolis, University of Minnesota Press, 1979, p. 15.

4. Ibid., p. 17; see also pp. 40 and 64.

5. Jean-François Lyotard and Jean-Loup Thébaud, *Just Gaming*, tr. Wlad Godzich, Minneapolis, University of Minnesota Press, 1985, e.g., pp. 93–94, 99.

6. Lyotard, *The Postmodern Condition*, pp. 15, 17. My interpolations.

7. Lyotard, *The Differend*, sec. 179. See also the analysis of the "economic genre" in sections 240ff., where Lyotard effectively equates the pragmatics of doings with those of sayings by analyzing exchange relations as linked phrases.

8. For further discussion of group ("social") space and the key role played in it by the distribution of economic capital, see Pierre Bourdieu, "Social Space and the Genesis of Groups," *Theory and Society* 14 (November 1985), pp. 723–744.

9. See my "Overdue Analysis of Bourdieu's Theory of Practice," *Inquiry* 30 (1987), pp. 113–135.

10. And in *Distinction: A Social Critique of the Judgement of Taste* (tr. Richard Nice, Cambridge, Mass., Harvard University Press, 1984), he writes that the fundamental oppositions at work in class-divided societies are those between "high/low, rich/poor etc." (p. 172; cf. pp. 468–469). This suggests that sets of fundamental oppositions might vary among (types of) society.

11. See also Pierre Bourdieu, *In Other Words: Essays toward a Reflexive Sociology*, tr. Matthew Adamson, Stanford, Calif., Stanford University Press, 1990, p. 65.

12. Bourdieu, *Distinction*, p. 468.

13. Concerning the anomalous status of this principle, see Hubert Dreyfus and Paul Rabinow, "Can There Be a Science of Existential Structure and Social Meaning? in *Bourdieu: Critical Perspectives*, ed. Craig Calhoun, Edward LiPuma, and Moishe Postone, Chicago, University of Chicago Press, 1993, pp. 35–44.

14. For extensive discussion of rules, see Bourdieu, *In Other Words: Essays toward a Reflexive Sociology*, chap. 3 and 4.

15. A similar point is made in Jacques Bouveresse's discussion of Bourdieu and Wittgenstein, though he does not point out how this problematizes Bourdieu's account (indeed, the very idea) of practical logic; Jacques Bouveresse, "Was ist eine Regel?" in *Praxis und Ästhetik. Neue Perspektiven im Denken Pierre Bourdieus*, ed. Gunter Gebauer and Christoph Wulf, Frankfurt am Main, Suhrkamp, 1994, p. 46. I might add that Taylor's assimilation of Bourdieu's habitus to the embodied understanding suggested in Wittgenstein rests on Bourdieu's critique of rules and does not confront the

conflict between practical logic and the inchoate nature of this understanding; Charles Taylor, "To Follow a Rule," in Calhoun, LiPuma, and Poston, *Bourdieu: Critical Perspectives*, pp. 45–60.

16. That is, in *Outline* and *Logic*. In other work, by contrast, Bourdieu seems to have taken his metatheoretical self-deflations more to heart. He no longer postulates a homologous structure to habitus and objective conditions, but instead searches for correlations between practices (life-styles) and such conditions (e.g., *Distinction*, part 2). The principles and structure of practical logic are mostly put aside, moreover, and habitus is primarily treated simply as "practical sense" or as a constellation of "senses" (e.g., *In Other Words*, chap. 3). To the extent that these changes have occurred, however, Bourdieu's account of practices indeed becomes vacuous: habitus becomes an empty intermediary between objective conditions and practices, i.e., life-styles (cf. *Distinction*, pp. 171, 467). Habitus becomes content-less insofar as it loses structure, i.e., practical logic. And insofar as it forfeits this content, it becomes functionally and essentially defined simply as that something, formed under certain conditions, that causes people to act sensibly in the games played under these conditions and that thereby produces practices that perpetuate the conditions. As a result, habitus also lacks explanatory power, since it explains neither why an actor, on any occasion, performs one sensible and condition-perpetuating action rather than another nor why people do anything at all that perpetuates these conditions. Perhaps for this reason Bourdieu continues to speak of "practical logic" and to describe dispositions, inter alia, as "the practical mastery of the logic or immanent necessity of a game" (*In Other Words*, p. 61). As we have seen, however, this "logic" and "necessity" cannot be adequately articulated. To speak of them is to gesture at the impossible content that Bourdieu needs to lend his account substance.

17. A similar point is made in William Sewell, Jr.'s discussion of Bourdieu and Giddens; see "A Theory of Structure: Duality, Agency, and Transformation, *American Journal of Sociology* 98 (1992), p. 11.

18. Lyotard, *The Differend*, p. xii.

19. See, e.g., Torsten Hägerstrand, "Space, Time, and Human Conditions," in *Dynamic Allocation of Urban Space*, ed. A. Karlqvist, L. Lundqvist, and F. Snickars, Lexington, Mass, Lexington Books, 1975, pp. 3–14.

20. Hubert Dreyfus, *Being-in-the-World: A Commentary on Heidegger's Being and Time, Division I*, Cambridge, Mass., MIT Press, 1991, pp. 189–191.

Chapter 6

1. E.g., middle Parsons's descriptions of a social system as an organization of interactions and of a social situation as a situation in which one actor is oriented toward another; Talcott Parsons and Edward Shils, *Toward a General Theory of Action*, New York, Harper and Row, 1962, pp. 7, 54–55, 88.

2. Construing interactions as the fundamental social category links theorists as varied as the phenomenologists Peter Berger and Thomas Luckmann, *The Social Construction of Reality*, Garden City, N.Y., Doubleday, 1966, p. 65; the presymbolic interactionist George Herbert Mead, *Mind, Self, and Society*, Chicago, University of Chicago Press, 1934; the ethnomethodologist Harold Garfinkel, *Studies in Ethnomethodology*, Englewood Cliffs, N.J., Prentice-Hall, 1967; and even, in some moods, Anthony Giddens, *Central Problems in Social Theory*, Berkeley, University of California Press, 1979, e.g., p. 66.

3. Note also that the basic unit of Habermas's communications theoretical reconstruction of the lifeworld is communicative interaction. See Jürgen Habermas, *The Theory of Communicative Action, Volume 2, Lifeworld and Systems: A Critique of Functionalist Reason*, tr. Thomas McCarthy, Boston, Beacon Press, 1987, chap. 6, sec. 1.

4. By a person's "life," I mean simply the manifold of actions a person performs along with the mental and cognitive conditions she is in. For the purposes of the present chapter, consequently, a person's life is coextensive with her mind/action; and because an individual is, above all, mind/action, the expressions "lives" and "individuals" will be used interchangeably.

5. For expansion of the brief sketch in this paragraph, see Alfred Schütz, *The Phenomenology of the Social World*, tr. George Walsh and Frederick Lehnert, Evanston, Ill., Northwestern University Press, 1967, sec. 29.

6. Ibid., p. 143. My translation.

7. Alfred Schütz and Thomas Luckmann, *Structures of the Lifeworld*, tr. Richard M. Zaner and H. Tristen Englehardt, Jr., Evanston, Ill., Northwestern University Press, 1973, part 4.

8. Schütz might have countered by claiming that social reality is identical with actors' experience and experiential knowledge of one another. See "Concept and Theory Formation in the Social Sciences," in *Collected Papers, Volume I*, The Hague, Martinus Nijhoff, 1962, p. 53; and Richard Grathoff, ed., *The Theory of Social Action*, Bloomington, Indiana University Press, 1978, p. 50. I will not engage such metaphysical claims in the present context.

9. See Ferdinand Tönnies, *Community and Association*, tr. Charles P. Loomis, London, Routledge and Kegan Paul, 1955. Tönnies's division builds upon earlier distinctions such as Hegel's contrast between *Sittlichkeit* and *Moralität*.

10. Max Weber, *Soziologische Grundbegriffe*, Tübingen, J. C. B. Mohr, 1960, p. 69; cf. Max Weber, *Basic Concepts of Sociology*, tr. H. P. Secher, Secaucus, N.J., Citadel, 1972, p. 91.

11. See Georg Simmel, "How Is Society Possible?" in *Philosophy of the Social Sciences*, ed. Maurice Natanson, New York, Random House, 1953, pp. 73–92; and Gerda Walther, "Zur Ontologie der sozialen Gemeinschaft," *Jahrbuch für Philosophie und phänomenologische Forschung* 6 (1923), p. 31.

12. The compositionist viewpoint persists, of course, to the present day. See, e.g., Margaret Gilbert's recent thesis, that a collection of individuals constitutes a social group if and only if each correctly thinks of himself and the others, taken together, as a we in a particular sense (*On Social Facts*, Princeton, Princeton University Press, 1989, chap. 4). Gilbert vacillates on the scope of her analysis. While admitting that human beings relate to one another in many ways having nothing to do with group membership (p. 223), she also asserts that her analysis of a group is at once an analysis of social phenomena generally (e.g., pp. 441–42). As she explains, this latter claim partly feeds off an intuitive equation of "social" (or *socius*) with ally. As intimated, I urge that social theory leave behind the now historically outdated Latin roots of this word and work with the modern, sociologically impacted, and wider equation of "social" with coexistence.

13. Oakeshott construes sociality as the association of "free agents" and distinguishes two basic forms of association: association by virtue of subscription to one and the same practice and association by virtue of partnership in the pursuit of a particular purpose. What marks both as forms of *Gesellschaft* is that belonging to an association of either type is something fundamentally up to individual choice. See Michael Oakeshott, *On Human Conduct*, Oxford, Clarendon Press, 1975, chap. 2, sec. 1–2.

14. Harold Blumer, *Symbolic Interactionism,* Englewood Cliffs, N.J., Prentice-Hall, 1969, pp. 38, 35, 7. Further references are in the text.

15. See H. Schmalenbach, "Die soziologische Kategorie des Bundes," *Dioskuren* 1 (1922), pp. 35–105, e.g., p. 50; and Alfred Vierkandt, *Gesellschaftslehre. Hauptprobleme der philosophischen Soziologie,* Stuttgart, Verlag von Ferdinand Enke, 1923, e.g., p. 28. Although Emile Durkheim and Wilhelm Dilthey do not operate explicitly in terms of the contrast between *Gemeinschaft* and *Gesellschaft,* important contrasts in their work (mechanical and organic solidarity in Durkheim and systems of culture and outer organizations of society in Dilthey) line up with Tönnies's distinction. While, however, Dilthey's analysis of social formations is to a large extent compositionist in spirit, Durkheim seeks to escape the compositionist framework. For Durkheim's account of solidarity, see Emile Durkheim, *The Division of Labor in Society,* tr. George Simpson, New York, Free Press, 1964; for his views on compositionism, see *The Rules of Sociological Method,* tr. Sarah A. Solovay and John H. Mueller, New York, Free Press, 1964. For Dilthey's analysis, see Wilhelm Dilthey, *Gesammelte Schriften,* Leipzig, B. G. Teubner, 1927, e.g., vol. 1, book 1, sec. 11–13.

16. Of course, this holds for modern, "demystified" thought alone. "Primitive" human beings often (implicitly) extended sociality to animals as well as inanimate objects and events.

17. Talcott Parsons, "Interaction," in *The International Encyclopedia of the Social Sciences,* ed. David L. Shills, New York, Macmillan, 1968, p. 440.

18. Niklas Luhmann, *Macht,* Opladen, Enke, 1975, p. 11; cf. Niklas Luhmann, "Einführende Bermerkungen zu einer Theorie symbolisch generalisierter Kommunikationsmedien," *Soziologische Aufklärung* 2, Opladen, Westdeutscher Verlag, 1975, pp. 170–192.

19. Habermas, *Theory of Communicative Action, Vol.* 2, e.g., pp. 181–185.

20. Tönnies, *Community and Association,* e.g., p. 48.

21. See Aron Gurwitsch, *Human Encounters in the Social World,* ed. Alexandre Métraux, tr. Fred Kersten, Pittsburgh, Duquesne University Press, 1977, p. 123.

22. Ibid., p. 127.

23. Ibid., p. 191 n. 171.

24. Wilhelm Dilthey and Count Paul Yorck von Wartenburg, *Briefwechsel zwischen Wilhelm Dilthey und dem Grafen Paul Yorck v. Wartenburg, 1877–1897,* Halle, Verlag Max Niemeyer, 1923, pp. 69, 71.

25. By "highly similar" practices I mean either "functionally" analogous ones, such as the language-games with the expressions "joy" and *Freude,* the dispersed practices of ordering and *Ordnen,* and multiple practices of X-ing; or "contentfully" highly similar ones such as rearing practices in Boston and Dallas.

26. See Weber, *Soziologische Grundbegriffe,* p. 54, cf. Weber, *Basic Concepts of Sociology,* p. 71; and George Homans, *The Nature of Social Science,* New York, Harcourt, Brace, and World, 1967, pp. 50–55.

27. Berger and Luckmann, *The Social Construction of Reality,* p. 65.

28. Jürgen Habermas, "A Review of Gadamer's *Truth and Method,*" in *Understanding and Social Inquiry,* ed. Fred R. Dallmayr and Thomas A. McCarthy, Notre Dame, Ind., University of Notre Dame Press, 1977, pp. 335–363.

29. See Hans-Georg Gadamer, "Rhetorik, Hermeneutik und Ideologiekritik," in *Hermeneutik und Ideologiekritik,* ed. Karl-Otto Apel, Frankfurt am Main, Suhrkamp, 1971, pp. 57–82.

30. Thus it also embraces other divisions in social phenomena based on the contrast between unreflective and reflective, e.g., Oakeshott's distinction between habitual and reflective morality: Michael Oakeshott, "The Tower of Babel," in his *Rationalism in Politics and Other Essays*, London, Methuen, 1962, pp. 59–79.

31. Talcott Parsons, *Sociological Theory and Modern Society*, New York, Free Press, 1967, p. 11.

32. I will simply accept these here undiscussed objections without comment. For a summary of criticisms pertinent to the current analysis, see Giddens, *Central Problems in Social Theory*, pp. 116–117.

33. Randall Collins, "On the Microfoundations of Macrosociology," *American Journal of Sociology* 86 (March 1981), pp. 984–1014.

34. Habermas, *Theory of Communicative Action, Volume 2*, p. 171.

35. Ibid., p. 267.

36. E.g., Nancy Hartsock, *Money, Sex, and Power: Towards a Feminist Historical Materialism*, New York, Longman, 1983.

37. Habermas claims that feedback mechanisms "functionally stabilize" as systems those clusters of actions whose unintended consequences so mesh as to fulfill certain general functions of society such as the "imperative" to maintain society's material substrate (*Theory of Communicative Action, Volume 2*, p. 232; cf. p. 151). He is silent, however, about exactly how feedback stabilizes actions as systems once their unintentional consequences so mesh as to have this effect. He may be right, for instance, that there is an imperative to maintain the material substrate in the sense that life is impossible if it is destroyed. But in this regard uncritically taking over Parsons's systems-functionalism, he fails to specify how it works that a cluster of actions becomes stabilized as a system through feedback when those of its unintended consequences that concern the transformation of this substrate combine to fulfill this imperative. Habermas rejects the idea that this feedback occurs by people recognizing what is occurring and acting intentionally to stabilize actions and their unintentional consequences. It is not clear, however, how else the fulfillment of the imperative can *through feedback*, and *over* chains of action mediated by intelligibility and physical connections, stabilize those actions whose unintentional consequences reproduce the substrate. Incidentally, this criticism parallels the widespread observation that functionalism fails to specify the causal mechanisms through which social formations bring about the accomplishments whose production is their "function." See, e.g., Jon Elster, *Explaining Technical Change*, Cambridge, Cambridge University Press, 1983, pp. 55–65.

38. Habermas, *Theory of Communicative Action, Volume 2*, p. 233.

39. See Marshall Berman, *All That Is Solid Melts into Air: The Experience of Modernity*, New York, Simon and Schuster, 1982.

40. Eugene Fink, *Existence und Coexistence*, Würzburg, Königshausen + Neumann, 1987, p. 128.

41. Gurwitsch, *Human Encounters in the Social World*, p. 127.

References

Adorno, Theodor. *Negative Dialectics*, tr. E. B. Ashton. New York, Continuum, 1973.

Anscombe, G. E. M., *Intention*. Ithaca, N.Y., Cornell University Press, 1957.

Baker, G. P., and P. M. S. Hacker. *Wittgenstein: Rules, Grammar and Necessity*. Oxford, Blackwell, 1986.

Bauman, Zygmunt. *Intimations of Postmodernity*. London, Routledge, 1992.

Bell, Catherine. *Ritual Theory, Ritual Practice*. Oxford, Oxford University Press, 1992.

Berger, Peter, and Thomas Luckmann. *The Social Construction of Reality*. Garden City, N.Y., Doubleday, 1966.

Berman, Marshall. *All That Is Solid Melts into Air: The Experience of Modernity*. New York, Simon and Schuster, 1982.

Black, Max. "*Lebensform* and *Sprachspiel* in Wittgenstein's Later Work." In *Wittgenstein and His Impact on Contemporary Thought: Proceedings of the Second International Wittgenstein Symposium*, ed. E. Leinfellner et al. Vienna, Verlag Hölder-Pichler-Tempsky, 1978, pp. 325–331.

Bloor, David. *Wittgenstein: A Social Theory of Knowledge*. New York, Columbia University Press, 1983.

Blumer, Harold. *Symbolic Interactionism*. Englewood Cliffs, N.J., Prentice-Hall, 1969.

Boss, Medard. *Existential Foundations of Medicine and Psychology*, tr. Stephen Conway and Anne Cleaves. New York, Jason Aronson, 1979.

Bourdieu, Pierre. *Outline of a Theory of Practice*, tr. Richard Nice. Cambridge, Cambridge University Press, 1976.

Distinction: A Social Critique of the Judgement of Taste, tr. Richard Nice. Cambridge, Mass., Harvard University Press, 1984.

"Social Space and the Genesis of Groups." *Theory and Society* 14 (November 1985), pp. 723–744.

The Logic of Practice, tr. Richard Nice. Stanford, Calif., Stanford University Press, 1990.

In Other Words: Essays toward a Reflexive Sociology, tr. Matthew Adamson. Stanford, Calif., Stanford University Press, 1990.

Bouveresse, Jacques. "Was ist eine Regel?" In *Praxis und Ästhetik. Neue Perspektiven im Denken Pierre Bourdieus*, ed. Gunter Gebauer and Christoph Wulf. Frankfurt am Main, Suhrkamp, 1993, pp. 41–56.

Budd, Malcolm. *Wittgenstein's Philosophy of Psychology*. New York, Routledge, 1989.

Burge, Tyler. "Individualism and the Mental." In *Midwest Studies in Philosophy IV: Studies in Metaphysics*, ed. Peter A. French, Theodore E. Uehling, and Howard K. Wettstein. Minneapolis, University of Minnesota Press, 1979, pp. 73–121.

Butler, Judith. *Gender Trouble: Feminism and the Subversion of Identity*. New York, Routledge, 1990.

Bodies That Matter: On the Discursive Limits of Sex. New York, Routledge, 1993.

Calhoun, Craig. "Culture, History, and the Specificity of Social Theory." In *Postmodernism and Social Theory*, ed. Steven Seidman and David G. Wagner. Oxford, Blackwell, 1992, pp. 244–288.

Churchland, Paul. "Eliminative Materialism and the Propositional Attitudes." *Journal of Philosophy* 78 (1981), pp. 67–90.

Collins, Randall. "On the Microfoundations of Macrosociology." *American Journal of Sociology* 86 (March 1981), pp. 984–1014.

Coulter, Jeff. *Mind in Action*. Atlantic Highlands, N.J., Humanities Press International, 1989.

Danto, Arthur. "Basic Actions." *American Philosophical Quarterly* 2, no. 2 (April 1965), pp. 141–148.

Davidson, Donald. "Mental Events." In *Experience and Theory*, ed. Lawrence Foster and J. W. Swanson. London, Duckworth, 1970, pp. 79–102.

"Belief and the Basis of Meaning." *Synthese* 27 (1974), pp. 309–323.

Day, Willard F. "On Certain Similarities between the *Philosophical Investigations* of Ludwig Wittgenstein and the Operationism of B. F. Skinner." *Journal of the Experimental Analysis of Behavior* 12, no. 3 (May 1969), pp. 489–506.

Dennett, Daniel. "Conditions of Personhood." In *The Identities of Persons*, ed. Amélie Oksenberg Rorty. Berkeley, University of California Press, 1976, pp. 175–196.

Dewey, John. *Human Nature and Conduct: An Introduction to Social Psychology*. London, George Allen and Unwin, 1922.

Dilthey, Wilhelm. *Gesammelte Schriften*, vol. 1 and 7. Leipzig, B. G. Teubner, 1927.

Dilthey, Wilhelm, and Count Paul Yorck von Wartenburg. *Briefwechsel zwischen Wilhelm Dilthey und dem Grafen Paul Yorck v. Wartenburg, 1877–1897*. Halle, Verlag Max Niemeyer, 1923.

Dreyfus, Hubert. *Being-in-the-World: A Commentary on Heidegger's Being and Time, Division I*. Cambridge, Mass., MIT Press, 1991.

Dreyfus, Hubert, and Stuart Dreyfus. *Mind over Machine*. New York, Free Press, 1986.

Dreyfus, Hubert, and Paul Rabinow. "Can There Be a Science of Existential Structure and Social Meaning?" In *Bourdieu: Critical Perspectives*, ed. Craig Calhoun, Edward LiPuma, and Moishe Postone. Chicago, University of Chicago Press, 1993, pp. 35–44.

Durkheim, Emile. *The Division of Labor in Society*, tr. George Simpson. New York, Free Press, 1964.

The Rules of Sociological Method, tr. Sarah A. Solovay and John H. Mueller. New York, Free Press, 1964.

Elster, Jon. *Explaining Technical Change: A Case Study in the Philosophy of Science*. Cambridge, Cambridge University Press, 1983.

The Cement of Society. Cambridge, Cambridge University Press, 1989.

Feyerabend, Paul. "Wittgenstein's 'Philosophical Investigations.'" *Philosophical Review* 64 (1955), pp. 449–483.

Finch, Henry Le Roy. *Wittgenstein – The Later Philosophy: An Exposition of the "Philosophical Investigations."* Atlantic Highlands, N.J., Humanities Press, 1977.

Fink, Eugene. *Existence und Coexistence*. Würzburg, Königshausen + Neumann, 1987.

Fleming, Noel. "Seeing the Soul." *Philosophy* 53 (1978), pp. 33–51.

Fodor, Jerry, and Charles Chihara. "Operationalism and Ordinary Language." In *Wittgenstein: The Philosophical Investigations*, ed. George Pitcher. London, Macmillan, 1968, pp. 384–419.

Foucault, Michel. "The History of Sexuality," tr. Leo Marshall. In *Power/Knowledge*, ed. Colin Gordon. New York, Pantheon, 1980, pp. 183–193.

——— *The History of Sexuality, Volume I*, tr. Robert Hurley. New York, Vintage, 1980, pp. 152–154.

——— "Two Lectures," tr. Kate Soper. In Michel Foucault, *Power/Knowledge*, ed. Colin Gordon. New York, Pantheon, 1980, pp. 78–108.

——— "The Subject and Power." In *Michel Foucault: Beyond Structuralism and Hermeneutics*, 2d ed., ed. Hubert L. Dreyfus and Paul Rabinow. Chicago, University of Chicago Press, 1983, pp. 208–226.

——— "Nietzsche, Genealogy, History." In *The Foucault Reader*, ed. Paul Rabinow. New York, Pantheon, 1984, pp. 76–100.

Gadamer, Hans-Georg. "Rhetorik, Hermeneutik und Ideologiekritik." In *Hermeneutik und Ideologiekritik*, ed. Karl-Otto Apel. Frankfurt am Main, Suhrkamp, 1971, pp. 57–82.

——— *Truth and Method*, 2d, rev. ed., tr. Joel Weinsheimer and Donald G. Marshall. New York, Crossroad, 1989.

Garfinkel, Harold. *Studies in Ethnomethodology*. Englewood Cliffs, N.J., Prentice-Hall, 1967.

Garver, Newton. "Form of Life in Wittgenstein's Later Works." *Dialectica* 44, nos. 1–2 (1990), pp. 175–201.

Giddens, Anthony. *Central Problems in Social Theory*. Berkeley, University of California Press, 1979.

——— *The Constitution of Society*. Berkeley, University of California Press, 1984.

——— "Time, Space, Regionalization." In *Social Relations and Spatial Structures*, ed. Derek Gregory and John Urry. London, Macmillan, 1985, pp. 265–295.

Gier, Nicholas F. *Wittgenstein and Phenomenology: A Comparative Study of the Later Wittgenstein, Husserl, Heidegger, and Merleau-Ponty*. Albany, State University of New York Press, 1981.

Gilbert, Margaret. *On Social Facts*. Princeton, Princeton University Press, 1989.

Goldman, Alvin. *A Theory of Human Action*. Princeton, Princeton University Press, 1970.

Grathoff Richard, ed. *The Theory of Social Action*. Bloomington, Indiana University Press, 1978.

Gregory, Derek. "Areal Differentiation and Post-Modern Human Geography." In *Horizons of Human Geography*, ed. Derek Gregory and Rex Walford. Totowa, N.J., Barnes and Noble, 1989, pp. 67–96.

Gurwitsch, Aron. *Human Encounters in the Social World*, ed. Alexandre Métraux, tr. Fred Kersten. Pittsburgh, Duquesne University Press, 1977.

Habermas, Jürgen. "A Review of Gadamer's *Truth and Method*." In *Understanding and Social Inquiry*, ed. Fred R. Dallmayr and Thomas A. McCarthy. Notre Dame, Ind., University of Notre Dame Press, 1977, pp. 335–363.

——— *The Theory of Communicative Action, Volume 2: Lifeworld and System: A Critique of Functionalist Reason*, tr. Thomas McCarthy. Boston, Beacon Press, 1987.

Hacker, P. M. S. *Insight and Illusion*, 2d ed., Oxford, Clarendon Press, 1986, chap. 3.

Hägerstrand, Torsten. "Space, Time, and Human Conditions," in *Dynamic*

Allocation of Urban Space, ed. A. Karlqvist, L. Lundqvist, and F. Snickars. Lexington, Mass., Lexington Books, 1975, pp. 3–14.

Haller, Rudolf. *Questions on Wittgenstein*. Lincoln, University of Nebraska Press, 1988.

Harré, Rom. *Physical Being*. Oxford, Blackwell, 1992.

Hartsock, Nancy. *Money, Sex, and Power: Towards a Feminist Historical Materialism*. New York, Longman, 1983.

Heidegger, Martin. "Letter on Humanism." In *Basic Writings*, ed. David Krell. New York, Harper and Row, 1977, pp. 193–242.

Being and Time, tr. John Macquarrie and Edward Robinson. Oxford, Blackwell, 1978.

Homans, George. *The Nature of Social Science*. New York, Harcourt, Brace, and World, 1967.

Hunter, J. F. M. "Forms of Life in Wittgenstein's Philosophical Investigations." *American Philosophical Quarterly* 5 (1968), pp. 233–243.

Janik, Allan. "Wie hat Schopenhauer Wittgenstein beeinflusst?" *Schopenhauer Jahrbuch* 73 (1992), pp. 69–77.

Joas, Hans. *Die Kreativität des Handelns*. Frankfurt am Main, Suhrkamp, 1992.

Johannessen, Kjell S. "Language, Art, and Aesthetic Practice." In *Wittgenstein – Aesthetics and Transcendental Philosophy*, ed. Kjell S. Johannessen and Tore Nordenstam. Vienna, Verlag Hölder-Pichler-Tempsky, 1981, pp. 108–126.

"Art, Philosophy and Intransitive Understanding." In *Wittgenstein – Eine Neubewertung*, ed. Rudolf Haller and Johannes Brandl. Vienna, Verlag Hölder-Pichler-Tempsky, 1990, pp. 323–333.

"Rule Following, Intransitive Understanding, and Tacit Knowledge." In *Essays in Pragmatic Philosophy*, ed. Helge Hoibraaten. Oxford, Oxford University Press, 1990, pp. 101–127.

Johnston, Paul. *Wittgenstein: Rethinking the Inner*. London, Routledge, 1993.

Khatchadourian, Haig. "Common Names and 'Family Resemblances.'" *Philosohy and Phenomenological Research*, 18 (1958), pp. 341–358.

Klages, Ludwig. *Vom Wesen des Bewusstseins*. Leipzig, Johann Ambrosius Barth, 1933.

Kristeva, Julia. *Revolution in Poetic Language*, tr. Margaret Waller. New York, Columbia University Press, 1984.

Lacan, Jacques. *Écrits: A Selection*, tr. Alan Sheridan. London, Tavistock, 1977.

Laclau, Ernesto, and Chantal Mouffe. *Hegemony and Socialist Strategy: Towards a Radical Democratic Politics*. London, Verso, 1985.

Louch, A. *Explanation and Human Action*. Berkeley, University of California Press, 1966.

Luhmann, Niklas. "Einführende Bermerkungen zu einer Theorie symbolisch generalisierter Kommunikationsmedien." *Soziologische Aufklärung 2*, Opladen, Westdeutscher Verlag, 1975, pp. 170–192.

Macht. Opladen, Enke, 1975.

Lynch, Michael. "Extending Wittgenstein: The Pivotal Move from Epistemology to the Sociology of Science." In *Science as Practice and Culture*, ed. Andrew Pickering. Chicago, University of Chicago Press, 1992, pp. 215–265.

Scientific Practice and Ordinary Action. Cambridge, Cambridge University Press, 1993.

Lyotard, Jean-François. *The Postmodern Condition: A Report on Knowledge*, tr.

Geoff Bennington and Brian Massumi. Minneapolis, University of Minnesota Press, 1979.
The Differend: Phrases in Dispute, tr. Georges van den Abbeele. Minneapolis, University of Minnesota Press, 1988.
Lyotard, Jean-François, and Jean-Loup Thébaud. *Just Gaming*, tr. Wlad Godzich. Minneapolis, University of Minnesota Press, 1988.
Malcolm, Norman. *Dreaming*. London, Routledge and Kegan Paul, 1962.
"Wittgenstein on the Nature of Mind." In *Thought and Knowledge*. Ithaca, N.Y., Cornell University Press, 1977, pp. 133–158.
"Wittgenstein on Language and Rules." *Philosophy* 64 (1989), pp. 5–28.
Mandelbaum, Maurice. "Social Facts." *British Journal of Sociology* 6 (1955), pp. 305–17.
Mann, Michael. *The Sources of Social Power, Volume I: A History of Power from the Beginning to A.D. 1760*. Cambridge, Cambridge University Press, 1986.
Mannheim, Karl. "Conservative Thought." In *From Karl Mannheim*, ed. Kurt H. Wolff. New York, Oxford University Press, 1971, pp. 132–222.
Mead, George Herbert. *Mind, Self, and Society*. Chicago, University of Chicago Press, 1934.
Melden, A. *Free Action*. London, Routledge and Kegan Paul, 1967.
Meløe, Jakob. "The Agent and His World." In *Praxeology: An Anthology*, ed. Gunnar Skirbekk. Oslo, Universitetsforlaget, 1983, pp. 13–29.
Merleau-Ponty, Maurice. *The Phenomenology of Perception*, rev. tr., tr. Colin Smith. London, Routledge, 1989.
Mouffe, Chantal. "Feminism, Citizenship, and Radical Democratic Politics." In *Feminists Theorize the Political*, ed. Judith Butler and Joan Scott. London, Routledge, 1992, pp. 369–384.
Mulhall, Stephen. *On Being in the World: Wittgenstein and Heidegger on Seeing Aspects*. New York, Routledge, 1990.
Nietzsche, Friedrich. *Beyond Good and Evil*, tr. Walter Kaufmann. New York, Random House, 1966.
The Twilight of the Idols. In *The Portable Nietzsche*, tr. Walter Kaufmann. New York, Penguin, 1982.
Nyiri, J. "Wittgenstein's Later Work in Relation to Conservatism." In *Wittgenstein and His Times*, ed. Brian McGuinness. Oxford, Blackwell, 1982, pp. 44–68.
Oakeshott, Michael, "Rational Conduct." In *Rationalism in Politics and Other Essays*. London, Methuen, 1962, pp. 80–110.
"The Tower of Babel." In *Rationalism in Politics and Other Essays*. London, Methuen, 1962, pp. 59–79.
On Human Conduct. Oxford, Clarendon Press, 1975.
Ortner, Sherry, "Theory in Anthropology since the Sixties." *Comparative Study of Society and History* 16 (1984), pp. 126–66.
Parsons, Talcott. *Sociological Theory and Modern Society*. New York, Free Press, 1967.
"Interaction." In *The International Encyclopedia of the Social Sciences*, ed. David L. Shills. New York, Macmillan, 1968.
The Structure of Social Action. New York, Free Press, 1968.
Parsons, Talcott, and Edward Shils. *Toward a General Theory of Action*. New York, Harper and Row, 1962.
Peacocke, Christopher. "Reply: Rule-Following: The Nature of Wittgenstein's Arguments." In *Wittgenstein: To Follow a Rule*, ed. S. H. Holtzman and C. M. Leich. London, Routledge and Kegan Paul, 1981, pp. 72–95.

Perby, Maja-Lisa. "Computerization and Skill in Local Weather Forecasting." In *Knowledge, Skill and Artificial Intelligence*, ed. Bo Göranzon and Ingela Josefson. Berlin, Springer-Verlag, 1988, pp. 39–52.

Plessner, Helmut. *Laughing and Crying: A Study of the Limits of Human Behavior*, tr. James Spencer Churchill and Marjorie Grene. Evanston, Ill., Northwestern University Press, 1970.

Quine, W. V. "Ontological Relativity." In *Ontological Relativity and Other Essays*. New York, Columbia University Press, 1969, pp. 26–68.

Robbins, Derek. *The Work of Pierre Bourdieu*. Milton Keynes, Open University Press, 1991.

Rorty, Richard. "Mind-Body Identity, Privacy, and the Categories." *Review of Metaphysics* 19, no. 1 (1965), pp. 24–54.

"Wittgensteinian Philosophy and Empirical Psychology." *Philosophical Studies* 31 (1977), pp. 151–172.

Philosophy and the Mirror of Nature. Princeton, Princeton University Press, 1979.

Sachs, David. "Wittgenstein on Emotion." *Acta Philosophica Fennica* 28 (1976), pp. 250–285.

Savigny, Eike von. "Avowals in the *Philosophical Investigations:* Expression, Reliability, Description." *Nous* 24 (1990), pp. 507–527.

"Self-Conscious Individual versus Social Soul: The Rationale of Wittgenstein's Discussion of Rule Following." *Philosophy and Phenomenological Research* 51, no. 1 (March 1991), pp. 67–84.

Der Mensch als Mitmensch. Studien zu Wittgensteins Philosophische Untersuchungen. Frankfurt am Main, Suhrkamp, 1995.

Schmalenbach, H. "Die soziologische Kategorie des Bundes," *Dioskuren* 1 (1922), pp. 35–105.

Schatzki, Theodore R. "Overdue Analysis of Bourdieu's Theory of Practice." *Inquiry* 30 (1987), pp. 113–35.

"Social Causality." *Inquiry* 31, no. 2 (1988), pp. 151–70.

"Elements of a Wittgensteinian Philosophy of the Human Sciences." *Synthese* 87 (1991), pp. 311–329.

"Spatial Ontology and Explanation." *Annals of the Association of American Geographers* 81, no. 4 (1991), pp. 650–670.

"Wittgenstein: Mind, Body, and Society." *Journal of the Theory of Social Behavior* 23, no. 3 (September 1993), pp. 285–314.

"Wittgenstein + Heidegger on the Stream of Life." *Inquiry* 36, no. 3 (September 1993), pp. 307–328.

Schütz, Alfred. *The Phenomentary of the Social World*, tr. George Walsh and Frederick Lehnert, Evanston, Ill., Northwestern University Press, 1967.

"Concept and Theory Formation in the Social Sciences." In *Collected Papers, Volume I*. The Hague, Martinus Nijhoff, 1962, pp. 48–66.

Schütz, Alfred, and Thomas Luckmann. *Structures of the Lifeworld*, tr. Richard M. Zaner and H. Tristen Englehardt, Jr. Evanston, Ill., Northwestern University Press, 1973.

Searle, John. *Speech Acts*. Cambridge, Cambridge University Press, 1969.

Seidman, Steven, and David G. Wagner, eds. *Postmodernism and Social Theory*, Oxford, Blackwell, 1992.

Sellin, Birger. *Ich will kein inmich mehr sein, botschaften aus einem autistischen kerker*, ed. Michael Klonovsky. Cologne, Kiepenheuer & Witsch, 1993.

Sewell, William, Jr. "A Theory of Structure: Duality, Agency, and Transformation." *American Journal of Sociology* 98 (1992), pp. 1–29.

Sheets-Johnstone, Maxine. *The Roots of Power: Animate Form and Gendered Bodies.* Chicago, Open Court, 1994.

Shoemaker, Sydney. "Self-Reference and Self-Awareness." *Journal of Philosophy* 65, no. 19 (October 1968), pp. 555–567.

Simmel, Georg. "How Is Society Possible?" In *Philosophy of the Social Sciences,* ed. Maurice Natanson. New York, Random House, 1953, pp. 73–92.

Soja, Edward. *Postmodern Geographies.* London, Verso, 1989.

Stern, David G. "Heraclitus' and Wittgenstein's River Images: Stepping Twice into the Same River." *Monist* 74, no. 4 (October 1991), pp. 581–604.

——— . *Wittgenstein on Mind and Language.* Oxford, Oxford University Press, 1994.

Strawson, Peter. *Individuals.* London, Methuen, 1959.

Taylor, Charles. "Interpretation and the Sciences of Man." In *Philosophy and the Human Sciences: Philosophical Papers 2.* Cambridge, Cambridge University Press, 1985, pp. 15–57.

——— . "To Follow a Rule . . ." In *Bourdieu: Critical Perspectives,* ed. Craig Calhoun, Edward LiPuma, and Moishe Postone. Chicago, University of Chicago Press, 1993, pp. 44–60.

Tilghman, Benjamin R. *But Is It Art? The Value of Art and the Temptation of Theory.* Oxford, Blackwell, 1984.

——— . *Wittgenstein, Ethics, and Aesthetics: The View from Eternity.* London, Macmillan, 1991.

Tönnies, Ferdinand. *Community and Association,* tr. Charles P. Loomis. London, Routledge and Kegan Paul, 1955.

Turner, Stephen. *The Social Theory of Practices: Tradition, Tacit Knowledge and Presuppositions.* Cambridge, Polity Press, 1994.

Vierkandt, Alfred. *Gesellschaftslehre. Hauptprobleme der philosophischen Soziologie.* Stuttgart, Verlag von Ferdinand Enke, 1923.

Waldenfels, Bernhard. *Ordnung im Zwielicht.* Frankfurt am Main, Suhrkamp, 1987.

Walther, Gerda. "Zur Ontologie der sozialen Gemeinschaft." *Jahrbuch für Philosophie und phänomenologische Forschung* 6 (1923), pp. 1–158.

Weber, Max. *Soziologische Grundbegriffe.* Tübingen, J. C. B. Mohr, 1960.

——— . *Basic Concepts of Sociology,* tr. H. P. Secher. Secaucus, N.J., Citadel, 1972.

Winch, Peter. *The Idea of a Social Science and Its Relation to Philosophy.* London, Routledge and Kegan Paul, 1958.

Wittgenstein, Ludwig. *Philosophical Investigations,* tr. G. E. M. Anscombe. New York, Macmillan, 1958.

——— . *Lectures and Conversations on Aesthetics, Psychology, and Religious Belief,* ed. Cyril Barrett. Oxford, Blackwell, 1966.

——— . *Zettel,* ed. G. E. M. Anscombe and G. H. von Wright, tr. G. E. M. Anscombe. Berkeley, University of California Press, 1967.

——— . *On Certainty,* tr. Denis Paul and G. E. M. Anscombe. Oxford, Blackwell, 1977.

——— . *Remarks on the Foundations of Mathematics,* rev. ed., ed. G. H. von Wright, R. Rhees, and G. E. M. Anscombe, tr. G. E. M. Anscombe. Cambridge, Mass., MIT Press, 1978.

——— . *Remarks on the Philosophy of Psychology,* vol. 1, G. E. M. Anscombe and G. H. von Wright, tr. G. E. M. Anscombe. Oxford, Blackwell, 1980.

——— . *Remarks on the Philosophy of Psychology,* vol. 2, ed. G. H. von Wright and Heikki Nyman, tr. C. G. Luckhardt and M. A. E. Aue. Oxford, Blackwell, 1980.

——— . *Last Writings on the Philosophy of Psychology, Volume I: Preliminary Studies for*

Part II of Philosophical Investigations, ed. G. H. von Wright and Heikki Nyman, tr. C. J. Luckhardt and M. A. E. Aue. Chicago, University of Chicago Press, 1990.

Last Writings on the Philosophy of Psychology, Volume II: The Inner and the Outer, ed. G. H. von Wright and Heikki Nyman, tr. C. G. Luckhardt and Maximilian A. E. Aue. Oxford, Blackwell, 1992.

"Notes for the Philosophical Lecture." In Ludwig Wittgenstein, *Philosophical Occasions,* ed. James Klagge and Alfred Nordman, tr. David G. Stern. Indianapolis, Ind., Hackett, 1993, pp. 445–457.

Wright, Georg Henrik von. *Explanation and Understanding.* Ithaca, N.Y., Cornell University Press, 1971.

"Wittgenstein in Relation to His Times." In *Wittgenstein and His Times,* ed. Brian McGuinness. Oxford, Blackwell, 1982, pp. 108–120.

Index

abidingness (Heidegger), 25
abilities (*see also* understanding):
knowing-how, 50, 91, 100, 103; open possibilities for action, 162; related to formulations, 49–51, 156–8
acceptability, *see* normativity
action intelligibility: applies to intended and unintended actions, 121; articulation of, 118; in Bourdieu, 153; not rationality, 118; practical logic (Bourdieu), 140; practices articulate, 116, 124; structuring of, 192–3
actions (*see also* behavior; chains of action): bodily doings and sensations in, 38–9; components of practices, 113; as conditions of life, 38–9; as continuous happening, 90; coordination of (Habermas), 89–90; fields of possible action, 159–67; governance of unreflective, 165–6; governed by teleoaffectivity, 123–4; joint actions (Blumer), 178–9; Oakeshott's account of, 120–1; Oakeshott's conception of, 96–8; presuppose practices, 92–3; reciprocal (*Wechselwirkungen*), 179; relation of habitus to (Bourdieu), 139–41; rules and resources as determinants of, 144–7; signifying, 121–2; signifying terminates in, 45; in social ontology, 19; underlies language (Wittgenstein), 135; unintended consequences of, 90
Adorno, Theodor, 225n
Althusser, Louis, 2
Anscombe, G. E. M., 118–19, 121, 218n
Apel, Karl-Otto, 56
appearances, *see* manifestations; phenomena
Aristotle, 59
articulation: defined, 111; of intelligibility, 116, 118, 124, 126–7, 130; of meaning, 117–18; within social practices, 112
Austin, John, 49

Baker, G. P., 223n
Bauman, Zygmunt, 3, 16–17
behavior (*see also* actions; doings; reactions; sayings): bodily doings and sayings as, 47; evidence of organization of a practice in, 101; in-particular-circumstances (Wittgenstein), 61; intelligibility of, 121; life conditions expressed in, 37–8; practices underlie patterns of, 37; reaction as spontaneous (Wittgenstein), 58–9; as type of action, 38–9; unreflective, 165; Wittgenstein's conception of, 42, 58
Berger, Peter, 188
Bergson, Henri, 26, 27, 55
Berman, Marshall, 230n
Black, Max, 219n
Bloor, David, 13, 56
Blumer, Harold, 178, 215n
body (*see also* doings; sayings): bodily doings and sayings, 22, 41–2, 47–8; bodily reper-

toire, 45, 51–3, 58, 60–1, 69, 98, 122, 125; expresses life conditions, 41–6; expressive body as social product, 52, 70; maturation of child's (Wittgenstein), 60–3, 65; neurophysiology of, 33, 59–60, 65; performance of bodily sayings, 51–2; as seat of mentality, 34–5, 41, 53; signifying of, 44–5; social context of bodily expression, 52–3; in Wittgenstein's conception of mind, 24–5
bodyhood (Boss), 42
Boss, Medard, 42
Bourdieu, Pierre: on habitus, 160; practice theory of, 13, 136–44
Bouveresse, Jacques, 26n
Budd, Malcolm, 94–5
Buddha, the, 69
Burge, Tyler, 220n
Butler, Judith, 35, 46–7, 85–6

capital (Bourdieu), 143–4
Carnap, Rudolf, 55
causal explanations (*see also* intelligibility explanations): ascriptions of mentality do not provide, 33; of life phenomena, 60, 76; natural facts have, 64; not part of Wittgenstein's enterprise, 56, 64–5; as tool of science, 64–5
causality (*see also* chains of action; expression): of action, 139; in body, 65; of mentality, 33, 80; not causal, 106–7, 109–10, 119; relation to the body, 33, 41–2, 59, 65, 139; in science, 60; via sociality, 192
chains of action: among practices, 89, 103, 122–3, 192; defined, 89, 191; intelligibility mediates, 193–4; in interactions, 207–8; within a practice, 89–90, 103, 192; as sociality within a practice, 191–2, 195–6
Chihara, Charles, 78–80
Christ, Jesus, 69
Churchland, Paul, 215n
coexistence, human (*see also* practices; social field; sociality): across practices, 198; compositionalist interpretation of, 176–84; in Gurwitsch, 182–4; integrative practices as phenomena of, 104–5; is a *Zusammenhang* (hanging-together), 14–15; locatedness in webs of, 195–6; medium of, 179; within a practice, 186–95, 208; related to sociality, 14–15; Schütz's interpretation of, 175; and the social, 13–14
cognitive, or intellectual, conditions, 37–8, 114–20; difference from mental conditions, 43–4; not explicitly directed states, 119–20; unexpressed, 72
Collins, Randall, 199
commonalities: of coexistence in sociality, 186–90; of we's, 208
communalization (Weber), 177
communication, *see* medium of communication
communitarianism, 6